工业和信息化精品系列教材
网络技术

Network Technology

微课版

路由交换
技术与实践
第 2 版

刘道刚 董良新 ◎ 主编
由丽娟 刘海莺 张召洪 孙新丽 ◎ 副主编

人民邮电出版社
北京

图书在版编目（CIP）数据

路由交换技术与实践：微课版 / 刘道刚，董良新主编. -- 2版. -- 北京：人民邮电出版社，2024.7
工业和信息化精品系列教材. 网络技术
ISBN 978-7-115-64292-9

Ⅰ. ①路⋯ Ⅱ. ①刘⋯ ②董⋯ Ⅲ. ①计算机网络－路由选择－教材②计算机网络－信息交换机－教材 Ⅳ. ①TN915.05

中国国家版本馆CIP数据核字(2024)第082184号

内 容 提 要

本书详细介绍了常用的路由与交换技术，分为6个模块20个项目，内容包括网络路由的配置、网络交换的配置、网络访问控制的配置、网络地址转换的配置、网络优化与安全的配置和路由器的管理。模块内的每个项目均包含用户需求、知识梳理、方案设计和项目实施，在项目的末尾有项目小结和拓展训练，可帮助读者总结和巩固所学的内容。

本书可以作为高校计算机网络技术专业及相关专业"路由与交换技术"课程的教材，也可以作为全国职业院校技能大赛相关赛项的备赛资料，还可以作为广大网络工程技术人员和计算机网络技术爱好者的自学参考书。

◆ 主　编　刘道刚　董良新
　　副主编　由丽娟　刘海莺　张召洪　孙新丽
　　责任编辑　顾梦宇
　　责任印制　王　郁　焦志炜

◆ 人民邮电出版社出版发行　北京市丰台区成寿寺路11号
　　邮编 100164　电子邮件 315@ptpress.com.cn
　　网址 https://www.ptpress.com.cn
　　三河市君旺印务有限公司印刷

◆ 开本：787×1092　1/16
　　印张：13.5　　　　　　　　2024年7月第2版
　　字数：365千字　　　　　　2024年7月河北第1次印刷

定价：49.80元

读者服务热线：(010)81055256　印装质量热线：(010)81055316
反盗版热线：(010)81055315
广告经营许可证：京东市监广登字 20170147 号

前言

路由器和交换机是网络中的核心设备。路由与交换技术及路由器与交换机的配置在网络组建、网络设备调试和网络管理过程中起着重要作用，是保证网络互联互通和可靠运行的关键。作为网络工程技术人员，掌握路由与交换技术，对完成路由器与交换机的配置至关重要。

本书编者在企业调研的基础上，结合全国职业院校技能大赛"网络系统管理"赛项考核的技能点和高等职业院校学生的学习特点，以实际应用转化的项目为主线，按"教、学、做"一体化的指导思想选取并安排内容。本书为培养技能型人才提供了合适的教学材料与训练内容。读者在学习本书的过程中，不仅可以学习基础知识，还可以结合所学内容按要求完成动手项目。为了适应时代发展，编者在编写本书的过程中融入了党的二十大精神，突出对读者职业道德、工匠精神的培养，满足职业院校对"路由与交换技术"课程的教学需求。

本书的主要特点如下。

1. 融入职业教育新理念

本书以项目为载体，分为 6 个模块 20 个项目，引导读者独立分析问题并解决问题，提高读者网络组建、配置和调试的能力。在设计项目的过程中，融入学思元素，注重培养读者的担当精神、遵纪守法的意识，以及质量强国和绿色发展的理念等，实现教学与育人的同向同行。

2. 理论与实践紧密结合

本书各项目均有对相关知识点、配置命令及其语法格式的介绍，在分析用户需求的基础上给出完成项目的解决方案，并配有详细的项目实施步骤。除此之外，每个项目都设置项目小结和拓展训练。本书将理论知识讲解与实践应用训练相结合，可提高读者的专业技能，有助于"教、学、做"一体化教学模式的实施。

3. 与技能大赛和行业对接

本书编者常年工作在教学一线，有着丰富的授课经验，多年来辅导学生参加各类职业技能大赛，参与了多种类型的教育教学改革与研究工作。本书编者在调研企业应用的基础上，深入分析了全国职业院校技能大赛考核的技能点，在内容组织上兼顾了行业认证的需求，有利于推动教学与技能大赛和行业的对接。

本书由福建中锐网络股份有限公司参与共同开发，在编写本书的过程中，编者得到了同事们和学生们的支持，以及来自思科网络技术学院和锐捷大学的网络工程技术人员的指导性意见，编者在此表示感谢。

由于编者水平有限，书中难免存在不妥之处，殷切希望广大读者批评指正，编者将不胜感激（编者 E-mail: yzlwldg@163.com）。

编者
2024 年 2 月

本书使用的图标

图标	设备名称	图标	设备名称
	路由器		集线器
	二层交换机		计算机
	三层交换机		服务器
	以太网连接		串行连接
	网络云	—	—

目录

模块一　网络路由的配置

项目 1

路由器的基本配置 ············· 2

- 1.1　用户需求 ····················· 2
- 1.2　知识梳理 ····················· 2
  - 1.2.1　认识路由器 ············· 2
  - 1.2.2　命令行界面的操作模式 ······· 3
  - 1.2.3　路由器的登录方式 ········· 4
  - 1.2.4　路由器的配置文件 ········· 4
  - 1.2.5　配置命令 ··············· 4
  - 1.2.6　快捷键 ················· 6
- 1.3　方案设计 ····················· 6
- 1.4　项目实施 ····················· 6
  - 1.4.1　路由器的 Console 口登录 ········· 6
  - 1.4.2　路由器的加密配置 ········· 7
  - 1.4.3　路由器以太口和计算机 IP 地址的配置 ············· 9
  - 1.4.4　路由器串口的配置 ········ 12
  - 1.4.5　路由器 telnet 的配置 ······ 12
- 1.5　项目小结 ···················· 14
- 1.6　拓展训练 ···················· 14

项目 2

静态路由的配置 ············· 15

- 2.1　用户需求 ···················· 15

- 2.2　知识梳理 ···················· 15
  - 2.2.1　直连路由 ··············· 15
  - 2.2.2　静态路由 ··············· 16
  - 2.2.3　常见静态路由介绍 ········ 17
- 2.3　方案设计 ···················· 18
- 2.4　项目实施 ···················· 18
  - 2.4.1　路由器接口和直连静态路由的配置 ··················· 18
  - 2.4.2　下一跳静态路由的配置 ····· 21
  - 2.4.3　完全指定静态路由的配置 ··· 23
  - 2.4.4　默认路由的配置 ·········· 25
  - 2.4.5　汇总静态路由的配置 ······ 27
  - 2.4.6　浮动静态路由的配置 ······ 29
- 2.5　项目小结 ···················· 33
- 2.6　拓展训练 ···················· 33

项目 3

RIP 的配置 ················· 34

- 3.1　用户需求 ···················· 34
- 3.2　知识梳理 ···················· 34
  - 3.2.1　动态路由协议及其分类 ····· 34
  - 3.2.2　RIP 的特点 ············· 35
  - 3.2.3　被动接口 ··············· 36
  - 3.2.4　RIP 的配置 ············· 36
  - 3.2.5　路由环路的避免 ·········· 37
  - 3.2.6　路由表 ················· 38
- 3.3　方案设计 ···················· 40
- 3.4　项目实施 ···················· 40

3.4.1　RIPv1 的配置……………40
　　3.4.2　RIPv1 不连续子网的配置………42
　　3.4.3　RIPv2 的配置……………44
　　3.4.4　被动接口的配置……………47
3.5　项目小结……………………47
3.6　拓展训练……………………47

项目 4

EIGRP 的配置……………48

4.1　用户需求……………………48
4.2　知识梳理……………………48
　　4.2.1　EIGRP 的特点……………48
　　4.2.2　与 EIGRP 相关的术语………49
　　4.2.3　EIGRP 的数据包类型………50
　　4.2.4　EIGRP 的度量值……………50
　　4.2.5　配置命令……………………50
4.3　方案设计……………………51
4.4　项目实施……………………51
　　4.4.1　一般情况下 EIGRP 的配置………51
　　4.4.2　禁用自动汇总时 EIGRP 的配置……………………53
4.5　项目小结……………………57
4.6　拓展训练……………………57

项目 5

OSPF 的配置……………59

5.1　用户需求……………………59
5.2　知识梳理……………………59
　　5.2.1　OSPF 的特点………………59
　　5.2.2　OSPF 的数据包类型………60
　　5.2.3　OSPF Router ID……………60
　　5.2.4　OSPF 的网络类型…………61

　　5.2.5　DR 和 BDR…………………61
　　5.2.6　虚链路………………………62
　　5.2.7　度量值………………………62
　　5.2.8　配置命令……………………62
5.3　方案设计……………………64
5.4　项目实施……………………64
　　5.4.1　单区域 OSPF 的配置………64
　　5.4.2　多区域 OSPF 的配置………66
　　5.4.3　骨干区域未与非骨干区域直连…68
　　5.4.4　OSPF 虚链路的配置………71
　　5.4.5　不同进程 OSPF 的配置……74
5.5　项目小结……………………76
5.6　拓展训练……………………76

项目 6

路由重分布……………77

6.1　用户需求……………………77
6.2　知识梳理……………………77
　　6.2.1　路由重分布的概念…………77
　　6.2.2　路由重分布的种类…………78
　　6.2.3　种子度量值…………………78
　　6.2.4　路由重分布的配置…………78
6.3　方案设计……………………79
6.4　项目实施……………………79
　　6.4.1　将静态路由重分布到 RIP……79
　　6.4.2　将静态路由重分布到 OSPF……82
　　6.4.3　RIP 和 OSPF 路由重分布的配置……………………85
　　6.4.4　不同进程 OSPF 路由重分布的配置……………………88
6.5　项目小结……………………92
6.6　拓展训练……………………92

模块二　网络交换的配置

项目 7

VLAN 的配置 ……………… 94

- 7.1 用户需求 ……………………… 94
- 7.2 知识梳理 ……………………… 94
  - 7.2.1 VLAN 的基本概念 ……… 94
  - 7.2.2 VLAN 的优点 …………… 94
  - 7.2.3 VLAN 的分类 …………… 95
  - 7.2.4 交换机的端口模式 ……… 96
  - 7.2.5 配置命令 ………………… 96
- 7.3 方案设计 ……………………… 97
- 7.4 项目实施 ……………………… 97
  - 7.4.1 交换机 VLAN 的查看 …… 97
  - 7.4.2 交换机 VLAN 的配置 …… 98
- 7.5 项目小结 ……………………… 99
- 7.6 拓展训练 ……………………… 100

项目 8

VLAN 中继的配置 ………… 101

- 8.1 用户需求 ……………………… 101
- 8.2 知识梳理 ……………………… 101
  - 8.2.1 VLAN 中继 ……………… 101
  - 8.2.2 帧在中继链路上的转发 … 101
  - 8.2.3 中继模式 ………………… 102
  - 8.2.4 VLAN 中继的配置 ……… 102
- 8.3 方案设计 ……………………… 103
- 8.4 项目实施 ……………………… 103
- 8.5 项目小结 ……………………… 106
- 8.6 拓展训练 ……………………… 106

项目 9

VLAN 间路由的配置 ……… 107

- 9.1 用户需求 ……………………… 107
- 9.2 知识梳理 ……………………… 107
  - 9.2.1 VLAN 间路由 …………… 107
  - 9.2.2 传统方式的 VLAN 间路由 … 108
  - 9.2.3 单臂路由 ………………… 108
  - 9.2.4 配置命令 ………………… 108
- 9.3 方案设计 ……………………… 108
- 9.4 项目实施 ……………………… 109
  - 9.4.1 传统方式 VLAN 间路由的配置 …………………… 109
  - 9.4.2 单臂路由的配置 ………… 111
- 9.5 项目小结 ……………………… 112
- 9.6 拓展训练 ……………………… 112

项目 10

三层交换机 VLAN 间路由的配置 …………………………… 114

- 10.1 用户需求 …………………… 114
- 10.2 知识梳理 …………………… 114
  - 10.2.1 三层交换机 …………… 114
  - 10.2.2 三层交换机端口 ……… 115
  - 10.2.3 三层交换机的 VLAN 间路由 …………………… 115
  - 10.2.4 配置命令 ……………… 115
- 10.3 方案设计 …………………… 115
- 10.4 项目实施 …………………… 115
- 10.5 项目小结 …………………… 117
- 10.6 拓展训练 …………………… 117

模块三 网络访问控制的配置

项目 11

部署 ACL 限制网络访问范围 ·················· 119

- 11.1 用户需求 ························· 119
- 11.2 知识梳理 ························· 119
  - 11.2.1 ACL 的概念 ············· 119
  - 11.2.2 ACL 的用途 ············· 119
  - 11.2.3 ACL 的工作原理 ········· 120
  - 11.2.4 ACL 的类型 ············· 120
  - 11.2.5 通配符掩码 ············· 120
  - 11.2.6 通配符掩码关键字 ······· 121
  - 11.2.7 ACL 创建原则 ··········· 121
  - 11.2.8 标准 ACL 的放置位置 ···· 121
  - 11.2.9 配置命令 ··············· 121
  - 11.2.10 ACL 的编辑 ············ 122
- 11.3 方案设计 ························· 123
- 11.4 项目实施 ························· 123
  - 11.4.1 标准编号 ACL 的配置 ···· 123
  - 11.4.2 标准命名 ACL 的配置 ···· 125
- 11.5 项目小结 ························· 127
- 11.6 拓展训练 ························· 127

项目 12

部署 ACL 限制网络流量 ····· 129

- 12.1 用户需求 ························· 129
- 12.2 知识梳理 ························· 129
  - 12.2.1 扩展 ACL ··············· 129
  - 12.2.2 端口号 ················· 130
  - 12.2.3 扩展 ACL 的放置位置 ···· 130

- 12.2.4 基于时间的 ACL ········· 130
- 12.2.5 配置命令 ··············· 131
- 12.3 方案设计 ························· 132
- 12.4 项目实施 ························· 132
  - 12.4.1 扩展编号 ACL 的配置 ···· 132
  - 12.4.2 扩展命名 ACL 的配置 ···· 135
  - 12.4.3 基于时间的 ACL 的配置 ···· 136
- 12.5 项目小结 ························· 138
- 12.6 拓展训练 ························· 139

模块四 网络地址转换的配置

项目 13

静态 NAT 的配置 ············ 141

- 13.1 用户需求 ························· 141
- 13.2 知识梳理 ························· 141
  - 13.2.1 公有地址和私有地址 ····· 141
  - 13.2.2 网络地址转换 ··········· 142
  - 13.2.3 静态 NAT ··············· 142
  - 13.2.4 NAT 的术语 ············· 142
  - 13.2.5 端口转发 ··············· 143
  - 13.2.6 配置命令 ··············· 143
- 13.3 方案设计 ························· 144
- 13.4 项目实施 ························· 144
  - 13.4.1 静态 NAT 的配置 ········ 144
  - 13.4.2 端口转发的配置 ········· 145
- 13.5 项目小结 ························· 146
- 13.6 拓展训练 ························· 146

项目 14

动态 NAT 的配置 ············ 147

- 14.1 用户需求 ························· 147

14.2 知识梳理 ······147
　14.2.1 动态 NAT ······147
　14.2.2 配置命令 ······148
14.3 方案设计 ······148
14.4 项目实施 ······148
14.5 项目小结 ······149
14.6 拓展训练 ······150

项目 15

基于端口的 NAT 的配置 ······ 151

15.1 用户需求 ······151
15.2 知识梳理 ······151
　15.2.1 基于端口的 NAT ······151
　15.2.2 配置命令 ······151
15.3 方案设计 ······152
15.4 项目实施 ······152
　15.4.1 基于端口的 NAT 的配置（单个内部接口）······152
　15.4.2 基于端口的 NAT 的配置（多个内部接口）······153
15.5 项目小结 ······155
15.6 拓展训练 ······155

模块五　网络优化与安全的配置

项目 16

生成树的配置 ······ 158

16.1 用户需求 ······158
16.2 知识梳理 ······158
　16.2.1 网络冗余 ······158
　16.2.2 网络中的二层环路 ······159
　16.2.3 生成树协议 ······159
　16.2.4 STP 的类型 ······159
　16.2.5 STP 的算法 ······160
　16.2.6 STP 的端口角色 ······160
　16.2.7 网桥 ID ······160
　16.2.8 STP 的收敛 ······161
　16.2.9 路径成本 ······161
　16.2.10 端口状态 ······162
　16.2.11 RSTP 端口 ······162
　16.2.12 边缘端口 ······163
　16.2.13 BPDU 防护 ······163
　16.2.14 配置命令 ······163
16.3 方案设计 ······164
16.4 项目实施 ······164
　16.4.1 根桥和端口角色的查看 ······164
　16.4.2 STP 的配置 ······166
　16.4.3 边缘端口和 BPDU 防护的配置 ······167
16.5 项目小结 ······168
16.6 拓展训练 ······169

项目 17

链路聚合的配置 ······ 170

17.1 用户需求 ······170
17.2 知识梳理 ······170
　17.2.1 链路聚合 ······170
　17.2.2 EtherChannel 技术 ······171
　17.2.3 创建 EtherChannel 的两种协议 ······171
　17.2.4 配置原则 ······172
　17.2.5 配置命令 ······172
17.3 方案设计 ······173
17.4 项目实施 ······173
17.5 项目小结 ······177

17.6 拓展训练 ························· 177

项目 18

交换机端口安全的配置 ······ 178

18.1 用户需求 ························· 178
18.2 知识梳理 ························· 178
    18.2.1 端口安全 ··················· 178
    18.2.2 安全 MAC 地址类型 ········ 179
    18.2.3 安全违规模式 ·············· 179
    18.2.4 发生安全违规的条件 ······· 180
    18.2.5 配置命令 ··················· 180
18.3 方案设计 ························· 181
18.4 项目实施 ························· 181
    18.4.1 静态端口安全的配置 ······· 181
    18.4.2 动态端口安全的配置 ······· 184
    18.4.3 粘滞端口安全的配置 ······· 188
18.5 项目小结 ························· 191
18.6 拓展训练 ························· 192

模块六 路由器的管理

项目 19

路由器的密码恢复 ············ 194

19.1 用户需求 ························· 194
19.2 知识梳理 ························· 194
    19.2.1 配置寄存器 ················ 194
    19.2.2 路由器密码的恢复 ········· 194
19.3 方案设计 ························· 194
19.4 项目实施 ························· 194
19.5 项目小结 ························· 195
19.6 拓展训练 ························· 196

项目 20

路由器 IOS 的备份与恢复 ··· 197

20.1 用户需求 ························· 197
20.2 知识梳理 ························· 197
    20.2.1 IOS ·························· 197
    20.2.2 IOS 的存放位置 ············ 197
    20.2.3 TFTP ······················· 198
    20.2.4 Xmodem 协议 ·············· 198
    20.2.5 配置命令 ··················· 198
20.3 方案设计 ························· 198
20.4 项目实施 ························· 199
    20.4.1 备份 IOS 到 TFTP 服务器 ···199
    20.4.2 用 TFTP 服务器恢复路由器的 IOS ·························· 201
    20.4.3 使用 ROMmon 模式恢复 IOS ·························· 201
20.5 项目小结 ························· 205
20.6 拓展训练 ························· 206

模块一

网络路由的配置

路由器能够把不同的网络互联起来,实现不同网络之间数据包的转发,隔离广播域和冲突域。当数据包在网络中转发时,路由器可以使数据包从源设备沿着正确的路径到达目的设备,实现路由选择。路由器要找到数据包去往目的设备的路径,正确的路由配置是必不可少的。本模块主要介绍路由器的基本配置、静态路由的配置、RIP 的配置、EIGRP 的配置、OSPF 的配置和路由重分布等。

项目 1
路由器的基本配置

1.1 用户需求

本项目网络拓扑如图 1-1 所示。由于路由器 R1 出现故障，学校采购了一台新路由器，怎样完成路由器的基本配置？

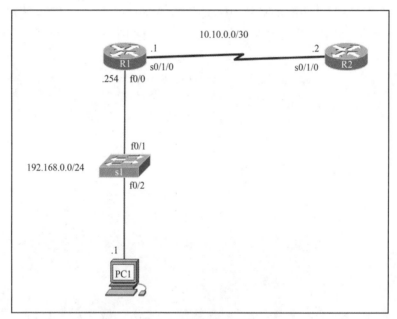

图 1-1 本项目网络拓扑

1.2 知识梳理

1.2.1 认识路由器

路由器实质上是一种特殊的计算机，是一种软硬件相结合的设备。路由器可以像计算机一样，执行操作系统指令，如系统初始化、路由和交换功能。路由器由中央处理器、随机存储器、只读存储器、闪存、非易失性随机访问存储器和操作系统等组成。

认识路由器

1. 中央处理器

中央处理器（Central Processing Unit，CPU）是路由器的核心。路由器的 CPU 与计算机的 CPU 功能相似，负责维护路由进程、运行路由算法、进行路由过滤和数据包转发等工作。CPU 的性能是衡量路由器性能的重要指标，路由器对数据包的处理速度很大程度上取决于 CPU 的性能。

2. 随机存储器

随机存储器（Random Access Memory，RAM）又称为易失性存储器，提供各种应用程序和进程的临时存储功能。RAM 中存储的内容有运行的操作系统、运行的配置文件、路由表项目、正在执行的代码和临时数据等，设备断电后，RAM 中的内容将会全部丢失。

3. 只读存储器

只读存储器（Read-Only Memory，ROM）又称为非易失性存储器，设备掉电后，其中的内容不会丢失。ROM 提供对启动说明、基本诊断软件和有限操作系统的永久存储功能。

4. 闪存

闪存是非易失性的，设备掉电后，其中的内容不会丢失。闪存提供 IOS 和其他相关系统文件的永久存储功能。

5. 非易失性随机访问存储器

非易失性随机访问存储器（Non-Volatile Random Access Memory，NVRAM）提供启动配置文件的永久存储功能，存储器掉电后，其中的内容不会丢失。

6. 操作系统

思科设备的操作系统（Internetwork Operating System，IOS）是用于网络设备操作和维护的操作系统，主要提供网络设备及连接端口的功能首选项设置、网络协议与网络功能设备之间数据传输的安全管理设置等功能。

1.2.2 命令行界面的操作模式

1. 用户执行模式

用户执行模式是系统启动后进入设备命令行界面（Command Line Interface，CLI）的第一种模式。该模式仅允许执行数量有限的基本监控命令，不允许执行任何可能改变设备配置的命令，通常称其为仅查看模式。

对于用户执行模式，采用以">"符号结尾的提示符标识，如下所示。

```
Router>
```

2. 特权执行模式

特权执行模式是管理员执行配置或管理命令时使用的模式。这个模式是进入全局配置模式和其他所有的具体配置模式之前必须通过的模式。对于特权执行模式，采用以"#"符号结尾的提示符标识，如下所示。

```
Router#
```

在用户执行模式下输入 enable 并按"Enter"键，可以从用户执行模式进入特权执行模式。在特权执行模式下输入 disable 或者 exit 并按"Enter"键，可以从特权执行模式退回用户执行模式。

3. 全局配置模式

全局配置模式又称为主配置模式，在全局配置模式下进行的命令行界面配置更改会影响设备的整体工作情况。在特权执行模式下输入 configure terminal，按"Enter"键后，提示符发生变化，表明路由器已处于全局配置模式，如下所示。

```
Router#configure terminal
Router(config)#
```
在全局配置模式下输入 exit，按 "Enter" 键后，可以从全局配置模式退回特权执行模式。

1.2.3 路由器的登录方式

1. Console 口登录方式

Console（控制台）口登录方式是指使用 Console 电缆把路由器和计算机连接起来，通过 Console 口对路由器进行带外访问。带外访问是指通过仅用于设备维护的专用管理通道进行访问。使用这种登录方式时，即使没有为设备配置任何网络服务，也可以访问设备。

2. 远程登录方式

远程登录方式采用通过虚拟接口在网络中建立远程设备命令行界面会话的方法，必须至少配置一个活动接口才可以使用这种登录方式。常用的远程登录方式有 telnet 和 SSH（Secure Shell）。

3. 辅助口登录方式

辅助（AUX）口登录方式使用连接到路由器辅助口的调制解调器进行拨号连接，类似于 Console 口登录方式。需要说明的是，辅助口登录方式也是带外访问。

路由器的登录方式

1.2.4 路由器的配置文件

1. 运行配置文件

运行配置文件是路由器当前配置状态的反映。其包含了应用于正在运行的设备的配置信息，通过 CLI 或者图形用户界面对设备进行的配置更改会立刻反映在配置文件上，并影响设备的实际运行状态。运行配置文件被存储在设备的内存或 RAM 中，如果设备断电或被重新启动，那么所有针对运行配置文件的未保存的更改都会丢失。

2. 启动配置文件

启动配置文件用于备份配置，它包含了设备在启动时需读取和应用的配置信息。启动配置文件被存储在 NVRAM 中。当配置了网络设备并修改了运行配置文件时，必须将这些更改保存到启动配置文件中，这样可防止所做的更改在电源发生故障或重新启动设备时丢失。

1.2.5 配置命令

1. 路由器的命名

```
Router(config)#hostname name
```

name：路由器的名称。路由器名称要符合以下要求：以字母开头，以字母或数字结尾，仅使用字母、数字和短横线连接号，不包含空格，长度小于 64 个字符。

2. 路由器的 Console 口加密

```
Router(config)#line console 0
Router(config-line)#password password
Router(config-line)#login
```

password：路由器的 Console 口登录密码。

login：对命令启用密码检查。

3. 使能密码的配置

```
Router(config)#enable password password
```

4. 使能加密密码的配置

```
Router(config)#enable secret password
```

5. 虚拟终端线路（Virtual Terminal Line，VTY）密码的配置

```
Router(config)#line vty 0 4
Router(config-line)#password password
Router(config-line)#login
```

6. 加密系统的所有明文密码

```
Router(config)#service password-encryption
```

7. 接口的配置

（1）指定接口的类型和编号以进行接口配置

```
Router(config)#interface type number
```

type number：接口的类型和编号。

（2）配置 IP 地址和子网掩码

```
Router(config-if)#ip address ip-address mask
```

ip-address：接口的 IP 地址。

mask：接口的 IP 地址对应的子网掩码。

（3）描述接口

```
Router(config-if)#description description
```

description：为每个接口配置的说明文字，说明文字的长度不能超过 240 个字符，可以在说明文字中说明接口所连接网络的信息，以及该网络中是否还有其他路由器等信息。

（4）激活接口

```
Router(config-if)#no shutdown
```

8. 保存配置

```
Router#copy running-config startup-config
Router#write memory
```

 注意 对于上述两条保存配置命令，只要在设备的特权执行模式下，配置其中的任意一条就可以实现对设备配置更改的保存。

9. 启动配置的删除

```
Router#erase startup-config
```

10. 设备的重启

```
Router#reload
```

11. 运行配置和启动配置的查看

```
Router#show running-config
Router#show startup-config
```

12. 与当前加载的 IOS 版本及硬件和设备相关的信息显示

```
Router#show version
```

13. 接口信息的查看

```
Router#show interfaces
Router#show ip interface brief
```

1.2.6 快捷键

"Tab"键用于自动补全当前正在输入的命令或关键字。
"Ctrl+Shift+6"组合键用于中断 ping 或 traceroute 等类型的 IOS 进程。
向下箭头键或"Ctrl+P"组合键用于向前滚动前面使用过的命令列表。
向上箭头键或"Ctrl+N"组合键用于向后滚动前面使用过的命令列表。
"Ctrl+C"组合键用于中断正在执行的命令或操作并退出配置模式。
"Ctrl+R"组合键用于重新显示一行。
"Ctrl+A"组合键用于将光标移至行首。
"Ctrl+Z"组合键用于退出全局配置模式并返回用户执行模式。

1.3 方案设计

路由器本身没有显示器等输出设备和键盘等输入设备,无法直接在其中输入命令和显示运行的参数。要完成路由器的配置,需要借助有输入设备和输出设备的计算机。因为是新买的路由器,没有为其配置任何网络服务,所以选择 Console 口登录方式。通过 Console 口登录后,完成路由器的加密配置,以及以太口、串口的配置和远程登录的配置等。

1.4 项目实施

1.4.1 路由器的 Console 口登录

本小节网络拓扑如图 1-1 所示,要求完成路由器 R1 的 Console 口登录的配置。
步骤 1:硬件连接。
Console 口登录的硬件连接如图 1-2 所示,将 Console 电缆的一端连接路由器的 Console 口,另一端连接计算机的 COM 口。

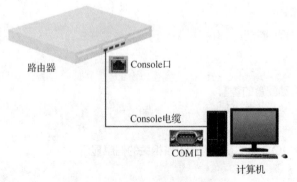

图 1-2　Console 口登录的硬件连接

步骤 2：在计算机中运行 PuTTY、Tera Term、SecureCRT 或者超级终端等软件，本项目运行的是 SecureCRT。

双击桌面上的 ▣（SecureCRT 的快捷方式）图标，打开 SecureCRT，单击 ▣ 图标（Quick Connect 图标），打开图 1-3 所示的"Quick Connect"对话框。

步骤 3：在"Protocol"下拉列表中选择"Serial"，在"Port"下拉列表中选择"COM1"，在"Baud rate"下拉列表中选择"9600"，取消勾选"RTS/CTS"复选框，如图 1-4 所示。

图 1-3　"Quick Connect"对话框 1　　　　图 1-4　"Quick Connect"对话框 2

步骤 4：单击"Connect"按钮，打开图 1-5 所示的窗口，计算机通过 Console 口成功登录路由器。

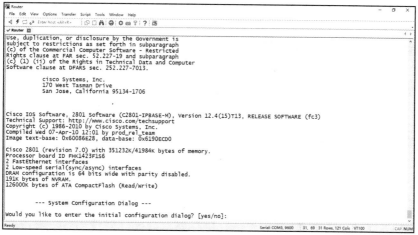

图 1-5　完成路由器的 Console 口登录

1.4.2　路由器的加密配置

本小节网络拓扑如图 1-1 所示，要求完成路由器 R1 的 Console 口的加密配置和保护特权执行模式的配置。

1. Console 口的加密配置

为网络设备的 Console 口配置密码是最低限度的安全措施，是在完成了路由器的 Console 口登录的基础上进行的。

步骤 1：在路由器 R1 的全局配置模式下配置 Console 口的加密，输入以下代码。

```
R1(config)#line console 0
R1(config-line)#password cisco
R1(config-line)#login
```

步骤2：输入多条 exit 命令退出用户执行模式，出现图 1-6 所示的提示。

```
R1 con0 is now available


Press RETURN to get started.


User Access Verification
Password:
```

图 1-6　Console 口加密的提示

步骤3：输入密码 cisco，进入用户执行模式。

2. 保护特权执行模式的配置

保护特权执行模式的安全性使用的是使能密码或者使能加密密码。

步骤1：在路由器 R1 的全局配置模式下完成使能密码的配置，输入以下代码。

```
R1(config)#enable password cisco
```

步骤2：从全局配置模式进入用户执行模式，在路由器 R1 的全局配置模式下输入以下代码。

```
R1(config)#exit
R1#disable
R1>enable
```

完成使能密码的配置后，从用户执行模式进入特权执行模式时会提示输入密码，如图 1-7 所示。

```
R1>enable
Password:
```

图 1-7　从用户执行模式进入特权执行模式时提示输入密码

步骤3：输入使能密码 cisco（输入后密码不显示），如图 1-8 所示。

```
R1>enable
Password:
R1#
```

图 1-8　输入使能密码

注意　输入密码时，路由器中不会显示"*"或者"#"，只要输入的密码正确，按"Enter"键后就可以进入特权执行模式。

步骤4：在路由器 R1 的全局配置模式下完成使能加密密码的配置，输入以下代码。

```
R1(config)#enable secret cisco1
```

重复执行步骤 2 和步骤 3，当从用户执行模式进入特权执行模式并提示输入密码时会发现，输入 cisco 无法进入特权执行模式，此时输入 cisco1 可以进入特权执行模式。

步骤 5：在路由器 R1 的特权执行模式下输入 show run | begin enable，查看运行配置文件，如图 1-9 所示。

```
R1#show run | begin enable
enable secret 5 $1$zhvn$dFNZvmDdmZRsNT462dxLb/
enable password cisco
```

图 1-9　查看运行配置文件

步骤 6：使路由器 R1 中密码的显示从明文方式变成密文方式，输入以下代码。

```
R1(config)#service password-encryption
```

输入密码时，通过让路由器 R1 从用户执行模式进入特权执行模式及查看路由器的运行配置文件等操作步骤会发现，当为一台路由器同时配置了使能密码和使能加密密码时，使能加密密码会起作用；在运行配置文件中，使能密码是以明文方式显示的，使能加密密码是以密文方式显示的，使能加密密码的安全性更高。

1.4.3　路由器以太口和计算机 IP 地址的配置

本小节网络拓扑如图 1-1 所示，要求完成路由器 R1 的以太口的配置和计算机 PC1 的 IP 地址的配置，实现计算机 PC1 与路由器 R1 的互联。

步骤 1：在路由器 R1 的全局配置模式下配置路由器 R1 的 f0/0 接口，输入以下代码。

```
R1(config)#interface fastEthernet 0/0
R1(config-if)#ip address 192.168.0.254 255.255.255.0
R1(config-if)#no shutdown
```

步骤 2：右击 PC1 桌面的网络图标，在弹出的快捷菜单中选择"属性"命令，如图 1-10 所示。

图 1-10　快捷菜单

步骤 3：选择"属性"命令后，打开"网络和共享中心"窗口，如图 1-11 所示。

图 1-11　"网络和共享中心"窗口

步骤 4:单击"查看活动网络"中的"以太网",打开"以太网 状态"对话框,如图 1-12 所示。

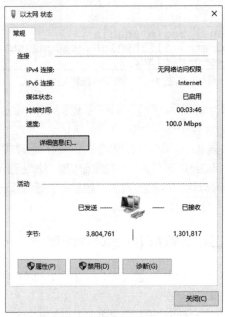

图 1-12 "以太网 状态"对话框

步骤 5:单击"属性"按钮,弹出"以太网 属性"对话框,如图 1-13 所示。

图 1-13 "以太网 属性"对话框

步骤 6:选择"Internet 协议版本 4(TCP/IPv4)"复选框,单击"属性"按钮,弹出"Internet 协议版本 4(TCP/IPv4)属性"对话框,如图 1-14 所示。

图 1-14 "Internet 协议版本 4(TCP/IPv4)属性"对话框

步骤 7：选中"使用下面的 IP 地址"单选按钮，根据图 1-1 中标注的计算机的 IP 地址完成配置，在"IP 地址："文本框中输入"192.168.0.1"，将"子网掩码："由网络前缀/24 转换成"255.255.255.0"，将"默认网关："配置为与该计算机直连的路由器接口的 IP 地址（路由器 R1 的 f0/0 接口的 IP 地址）"192.168.0.254"，如图 1-15 所示。

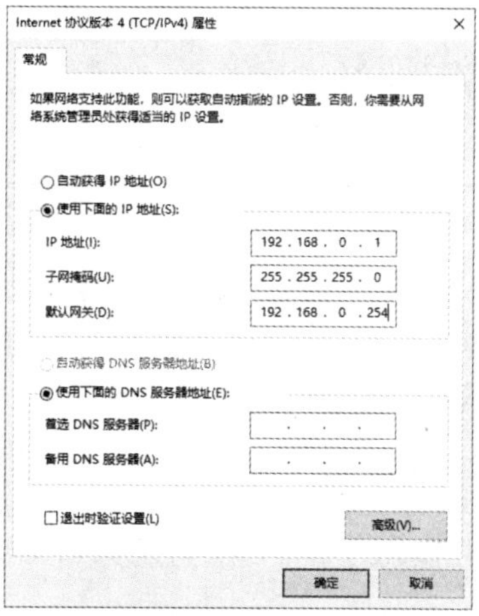

图 1-15 配置 IPv4 地址

步骤 8：单击"确定"按钮，如果没有提示 IP 地址冲突，则说明配置成功。
步骤 9：检验联通性，从计算机 PC1 ping 路由器 R1，如图 1-16 所示。

```
C:\>ping 192.168.0.254

正在 Ping 192.168.0.254 具有 32 字节的数据:
来自 192.168.0.254 的回复: 字节=32 时间<1ms TTL=255
来自 192.168.0.254 的回复: 字节=32 时间<1ms TTL=255
来自 192.168.0.254 的回复: 字节=32 时间<1ms TTL=255
来自 192.168.0.254 的回复: 字节=32 时间<1ms TTL=255

192.168.0.254 的 Ping 统计信息:
    数据包: 已发送 = 4，已接收 = 4，丢失 = 0 (0% 丢失)，
往返行程的估计时间(以毫秒为单位):
    最短 = 0ms，最长 = 0ms，平均 = 0ms
```

图 1-16 从计算机 PC1 ping 路由器 R1

通过查看从计算机 PC1 ping 路由器 R1 返回的结果可以发现，计算机 PC1 和路由器 R1 已经实现了互通。

1.4.4 路由器串口的配置

本小节网络拓扑如图 1-1 所示，要求完成路由器 R1 和 R2 串口的配置，实现路由器 R1 和 R2 的互通。

步骤 1：在路由器 R1 的全局配置模式下配置路由器 R1 的 s0/1/0 接口，设置 IP 地址及子网掩码，输入以下代码。

```
R1(config)#interface serial 0/1/0
R1(config-if)#ip address 10.10.0.1 255.255.255.252
R1(config-if)#no shutdown
```

步骤 2：在路由器 R2 的全局配置模式下配置路由器 R2 的 s0/1/0 接口，设置 IP 地址及子网掩码，输入以下代码。

```
R2(config)#interface serial 0/1/0
R2(config-if)#ip address 10.10.0.2 255.255.255.252
R2(config-if)#no shutdown
```

步骤 3：在路由器 R1 的特权执行模式下输入 ping 10.10.0.2，检验联通性，如图 1-17 所示。

```
R1#ping 10.10.0.2

Type escape sequence to abort.
Sending 5, 100-byte ICMP Echos to 10.10.0.2, timeout is 2 seconds:
!!!!!
Success rate is 100 percent (5/5), round-trip min/avg/max = 12/15/16 ms
```

图 1-17　从路由器 R1 ping 路由器 R2

从路由器或者交换机等网络设备发出的 ping 命令将会为发送的每个互联网控制报文协议（Internet Control Message Protocol，ICMP）应答并生成一个指示符，常见的指示符如下。

"!"：表示收到一个 ICMP 应答。

"."：表示等待应答时超时。

"U"：表示收到了一个 ICMP 无法到达报文。

通过查看从路由器 R1 ping 路由器 R2 返回的结果可以发现，路由器 R1 和 R2 已经实现了互通。

1.4.5 路由器 telnet 的配置

1. telnet 时不进行身份验证

本小节网络拓扑如图 1-1 所示，要求完成路由器 R2 的 telnet 的配置，实现从路由器 R1 telnet 到路由器 R2，telnet 时不进行身份验证。

步骤 1：在路由器 R2 的全局配置模式下配置使能加密密码和 telnet，输入以下代码。

```
R2(config)#enable secret 123
R2(config)#line vty 0 4
R2(config-line)#no login
```

步骤 2：在路由器 R1 的特权执行模式下输入 telnet 10.10.0.2，进入路由器 R2 的用户执行模式，如图 1-18 所示。

```
R1#telnet 10.10.0.2
Trying 10.10.0.2 ... Open
R2>
```

图 1-18　从路由器 R1 telnet 路由器 R2

通过查看从路由器 R1 telnet 路由器 R2 的状态可以发现，从路由器 R1 成功远程登录路由器 R2。

2. 远程登录时进行密码验证

本小节网络拓扑如图 1-1 所示，要求完成路由器 R2 的 telnet 的配置，实现从路由器 R1 远程登录路由器 R2，telnet 时要验证密码。

步骤 1：在路由器 R2 的全局配置模式下配置使能加密密码和 telnet，输入以下代码。

```
R2(config)#enable secret 123
R2(config)#line vty 0 4
R2(config-line)#password 123
R2(config-line)#login
```

步骤 2：在路由器 R1 的特权执行模式下输入 telnet 10.10.0.2，提示输入密码，这里输入密码 123，进入路由器 R2 的用户执行模式，如图 1-19 所示。

```
R1#telnet 10.10.0.2
Trying 10.10.0.2 ... Open

User Access Verification

Password:
R2>
```

图 1-19　从路由器 R1 telnet 路由器 R2

通过查看从路由器 R1 telnet 路由器 R2 的状态可以发现，从路由器 R1 成功 telnet 路由器 R2，telnet 时验证了密码。

3. telnet 时验证用户名和密码

本小节网络拓扑如图 1-1 所示，要求完成路由器 R2 的 telnet 的配置，实现从路由器 R1 telnet 路由器 R2，telnet 时验证用户名和密码。

步骤 1：在路由器 R2 的全局配置模式下配置使能加密密码和 telnet，输入以下代码。

```
R2(config)#enable secret 123
R2(config)#username R2 password 123
R2(config)#line vty 0 4
R2(config-line)#login local
```

步骤 2：在路由器 R1 的特权执行模式下输入 telnet 10.10.0.2，提示输入用户名和密码，这里输入用户名 R2 和密码 123，进入路由器 R2 的用户执行模式，如图 1-20 所示。

```
R1#telnet 10.10.0.2
Trying 10.10.0.2 ... Open

User Access Verification

Username: R2
Password:
```

图 1-20　进入路由器 R2 的用户执行模式

通过查看从路由器 R1 telnet 路由器 R2 的状态可以发现，从路由器 R1 成功 telnet 路由器 R2，telnet 时验证了用户名和密码。

1.5 项目小结

本项目首先完成了路由器的 Console 口登录，并在此基础上完成了路由器的加密配置、路由器以太口和计算机 IP 地址的配置、路由器串口的配置，以及路由器 telnet 的配置等，这些配置是新路由器需进行的基本配置。

> 🔍 边学边思
> 熟练掌握路由器的基本配置对学好"路由与交换技术"课程非常重要。打好基础是取得成功的关键。无论是学习、职业发展还是个人成长，将基础知识和技能学扎实都是持久而有益的投资。

1.6 拓展训练

本项目拓展训练网络拓扑如图 1-21 所示，要求完成如下配置：
（1）完成路由器 R1 的接口和计算机 PC1 与 PC2 的 IP 地址的配置，实现网络互通；
（2）完成路由器 R1 的 Console 口的加密配置；
（3）完成路由器 R1 的 telnet 的配置，配置 telnet 时要验证用户名和密码，并且远程登录后能够对设备进行配置；
（4）配置过程中用到的用户名为 R1，密码为 cisco。

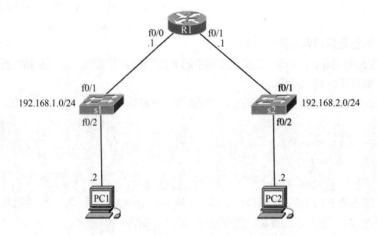

图 1-21　项目 1 拓展训练网络拓扑

项目 2
静态路由的配置

2.1 用户需求

本项目网络拓扑如图 2-1 所示，要求配置静态路由，实现计算机 PC1、PC2 和 PC3 互通。

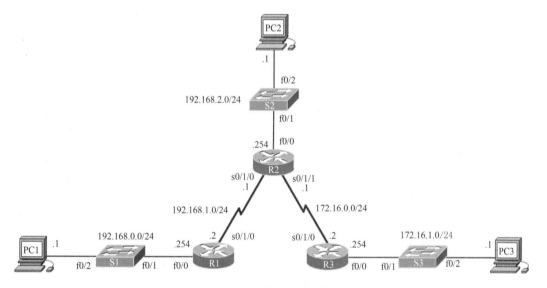

图 2-1 本项目网络拓扑

2.2 知识梳理

2.2.1 直连路由

直连路由

直连路由是路由器接口所连接子网的路由方式。当使用 no shutdown 命令将直连的其他设备的路由器接口激活后，该接口处于 up/up 状态，并且该接口也配置了 IP 地址。该接口所在的网络就会作为直连网络加入路由表，形成一条直连路由。

1. 直连路由出现在路由表中的条件

（1）接口处于 up/up 状态。
（2）已经完成了接口的 IP 地址的配置。

2. 直连路由的检查

用 show ip route 命令查看路由表，路由器 R1 的路由表如图 2-2 所示。

```
R1#show ip route
Codes: L - local, C - connected, S - static, R - RIP, M - mobile, B - BGP
       D - EIGRP, EX - EIGRP external, O - OSPF, IA - OSPF inter area
       N1 - OSPF NSSA external type 1, N2 - OSPF NSSA external type 2
       E1 - OSPF external type 1, E2 - OSPF external type 2
       i - IS-IS, su - IS-IS summary, L1 - IS-IS level-1, L2 - IS-IS level-2
       ia - IS-IS inter area, * - candidate default, U - per-user static route
       o - ODR, P - periodic downloaded static route, H - NHRP, l - LISP
       + - replicated route, % - next hop override

Gateway of last resort is not set

      192.168.0.0/24 is variably subnetted, 2 subnets, 2 masks
C        192.168.0.0/24 is directly connected, FastEthernet0/0
L        192.168.0.254/32 is directly connected, FastEthernet0/0
```

图 2-2　路由器 R1 的路由表

路由表中的第一项是路由来源，用于确定路由的获取方式。图 2-2 中，直连接口有两条路由：一条是路由来源为"C"的路由，"C"用于标识直连网络；另外一条是路由来源为"L"的路由，"L"用于标识为路由器接口分配的 IPv4 地址。使用不同 IOS 版本的路由器的直连条目是不同的。如果路由器用的是 IOS 15 之前的版本，那么路由表的直连条目中只有路由来源为"C"的直连路由，没有路由来源为"L"的本地路由。如果路由器用的是 IOS 15 之后的版本，那么直连条目中路由来源为"C"的直连路由和路由来源为"L"的本地路由都有。

直连路由是配置静态路由和动态路由的基础，在为路由器配置静态或动态路由之前，如果路由器的路由表中没有直连路由，那么静态路由和动态路由将不会在路由表中生成。

2.2.2　静态路由

静态路由是指由网络管理员手动配置的路由，用于定义去往目的网络的明确路径，这种路由一经写入就不会自动修改。若网络拓扑发生变化，则必须由网络管理员来手动修改。

静态路由

1. 静态路由的类型

根据静态路由的功能和特征，静态路由可分为标准静态路由、默认路由、汇总静态路由和浮动静态路由。

根据如何指定目标，静态路由又可以分为以下 3 种：仅指定下一跳 IP 地址的称为下一跳静态路由；仅指定路由器送出接口的称为直连静态路由；指定下一跳 IP 地址和送出接口的称为完全指定静态路由。

2. 静态路由的配置方法

（1）检查每台路由器中直连路由的条数。图 2-1 所示的网络拓扑中有 3 台路由器 R1、R2 和 R3，路由器 R1 直连了两个网络，应该有两条直连路由；路由器 R2 直连了 3 个网络，应该有 3 条直连路由；路由器 R3 直连了两个网络，应该有两条直连路由。

（2）分析网络拓扑，找到每台路由器需要配置的静态路由的条数。在网络中，如果要确保数据包能够成功到达目的网络，那么每个路由器的路由表都需要包含能够到达该目的网络的路由信息。除了直接连接到路由器的网络外，对于其他非直连的网络，路由器需要配置静态路由来指示数据包应该通过哪条路径到达目的网络。图 2-1 所示的网络拓扑中，3 台路由器 R1、R2 和 R3 连接了 5 个网络。当网络基本配置完成后，路由器 R1 的路由表中有两条直连路由，缺少去往 3 个（5-2）非直连网络的 3 条路由。要保证路由器 R1 能够把数据包发送到整个网络，路由器 R1 需要为 3 个非直连网络配置 3 条静态路由。同理，路由器 R2 需要配置两条静态路由，路由器 R3 需要配置 3 条静态路由。

（3）用静态路由配置命令为网络中的每台路由器配置相应的静态路由。

3. 静态路由的配置命令

```
Router(config)#ip route network-address subnet-mask {ip-address|
interface-type interface-number [ip-address]}[distance][name name]
[permanent][tag tag]
```

network-address：要加入路由表的远程目的网络的网络地址。

subnet-mask：要加入路由表的远程网络的子网掩码。可对此子网掩码进行修改，以汇总一组网络。

ip-address：将数据包转发到远程目的网络的下一跳路由器的 IP 地址，一般称为下一跳 IP 地址。

interface-type interface-number：用于将数据包转发到下一跳路由器的发送接口。

distance：可选参数，静态路由的管理距离，配置浮动静态路由时，可通过修改该参数实现路由的浮动。

name：可选参数，用于指定下一跳的名称。

permanent：可选参数，可将静态路由标记为永久路由。

tag：可选参数，用于给静态路由指定一个标签，以便识别和管理静态路由。

4. 查看路由表

```
Router#show ip route
```

2.2.3 常见静态路由介绍

1. 默认路由

默认路由是将 0.0.0.0/0 作为目的 IPv4 地址的静态路由，是静态路由的一种特殊形式。如果数据包能够与路由表中非默认路由匹配，那么路由器将不会查看默认路由。当数据包与路由表中非默认路由都不匹配时，路由器会查看默认路由。如果为路由器配置了默认路由，那么数据包一定能够与默认路由匹配，依据默认路由指示的路径被转发出去。作为未知地址数据包的最后处理方式，配置默认路由可以创建选用的网关。

默认路由、汇总静态路由和浮动静态路由

2. 汇总静态路由

汇总静态路由是多条静态路由汇总成的一条静态路由。若目的网络是连续的，则可以把连续的网络地址汇总成一个网络地址，并且多条静态路由都使用相同的送出接口或下一跳 IP 地址，这时可以通过把多个目的网络汇总成一个网络，把多条静态路由汇总成一条汇总静态路由。汇总静态路由可以减少路由表中条目的数量。

3. 浮动静态路由

在介绍浮动静态路由之前，首先介绍什么是管理距离。管理距离定义了路由来源的优先级别。对于每个路由来源（包括特定路由协议、静态路由或直连网络），管理距离值小的路由优先级高，管理距离值大的路由优先级低。如果从多个不同的路由来源获取同一目的网络的路由信息，那么路由器会使用管理距离来选择最佳路径。管理距离值是从 0～255 的整数，值越小表示路由来源的优先级越高。管理距离值为 0，表示优先级最高。

如果去往目的网络有多条路径，那么当网络正常运行时，所有数据包都从某一路径去往目的网络，当这条路径出现故障时，数据包会从其他路径去往目的网络。网络正常运行时数据包通过的这条路径称为主路径，其他路径称为备用路径。

浮动静态路由是为主路径静态或动态路由提供备用路径的静态路由,它仅在主路径不可用时使用。浮动静态路由的管理距离值比主路径路由的管理距离值大。当主路径和备用路径都可以到达目的网络时,路由器会选择管理距离值小的主路径;当主路径出现故障后,路由器会把管理距离值大于主路径路由的备用路径路由添加到路由表中,实现了路由的浮动。

> 🔍 边学边思
>
> 静态路由不会自动修改,这种"持之以恒"的特性,保证了路由器可以稳定地把数据包转发到目的设备。我们也应该培养持之以恒、勇于担当、吃苦耐劳、积极奋斗的精神,以求在学习和各项工作中取得优异成果。

2.3 方案设计

图 2-1 所示的网络拓扑中,3 台路由器 R1、R2 和 R3 连接了 5 个网络,路由器 R1 有 3 个非直连网络,路由器 R2 有两个非直连网络,路由器 R3 有 3 个非直连网络。因为路由器 R1、R2 和 R3 没有去往非直连网络的路由,所以网络无法互通。要实现网络的互通,可以针对路由器 R1、R2 和 R3 的非直连网络配置标准静态路由。

通过观察图 2-1 所示的网络拓扑,可以看出以下内容。

路由器 R1 要把数据包转发给 192.168.2.0/24、172.16.0.0/24 和 172.16.1.0/24 这 3 个非直连网络,都需要把数据包沿着同一路径转发给路由器 R2,所以可以针对这 3 个非直连网络,为路由器 R1 配置一条默认路由。也可以把 172.16.0.0/24 和 172.16.1.0/24 这两个网络汇总,配置一条汇总静态路由。

路由器 R3 要把数据包转发给 192.168.0.0/24、192.168.1.0/24 和 192.168.2.0/24 这 3 个非直连网络,都需要把数据包沿着同一路径转发给路由器 R1,所以可以针对这 3 个非直连网络,为路由器 R3 配置一条默认路由。也可以把 192.168.0.0/24、192.168.1.0/24 和 192.168.2.0/24 这 3 个网络汇总,配置一条汇总静态路由。

2.4 项目实施

2.4.1 路由器接口和直连静态路由的配置

本小节网络拓扑如图 2-1 所示,计算机 PC1、PC2 和 PC3 的 IP 地址已经配置完成,要求完成路由器接口的配置和直连静态路由的配置,实现计算机 PC1、PC2 和 PC3 互通。

步骤 1:在路由器 R1 的全局配置模式下配置接口,输入以下代码。

```
R1(config)#interface f0/0
R1(config-if)#ip address 192.168.0.254 255.255.255.0
R1(config-if)#no shutdown
R1(config-if)#interface s0/1/0
R1(config-if)#ip address 192.168.1.2 255.255.255.0
R1(config-if)#no shutdown
```

步骤 2:在路由器 R2 的全局配置模式下配置接口,输入以下代码。

```
R2(config)#interface f0/0
```

```
R2(config-if)#ip address 192.168.2.254 255.255.255.0
R2(config-if)#no shutdown
R2(config-if)#interface s0/1/0
R2(config-if)#ip address 192.168.1.1 255.255.255.0
R2(config-if)#no shutdown
R2(config-if)#interface s0/1/1
R2(config-if)#ip address 172.16.0.1 255.255.255.0
R2(config-if)#no shutdown
```

步骤 3：在路由器 R3 的全局配置模式下配置接口，输入以下代码。

```
R3(config)#interface f0/0
R3(config-if)#ip address 172.16.1.254 255.255.255.0
R3(config-if)#no shutdown
R3(config-if)#interface s0/1/0
R3(config-if)#ip address 172.16.0.2 255.255.255.0
R3(config-if)#no shutdown
```

步骤 4：在路由器 R1 的特权执行模式下输入 show ip route，查看路由表，如图 2-3 所示。

```
R1#show ip route
Codes: C - connected, S - static, R - RIP, M - mobile, B - BGP
       D - EIGRP, EX - EIGRP external, O - OSPF, IA - OSPF inter area
       N1 - OSPF NSSA external type 1, N2 - OSPF NSSA external type 2
       E1 - OSPF external type 1, E2 - OSPF external type 2
       i - IS-IS, su - IS-IS summary, L1 - IS-IS level-1, L2 - IS-IS level-2
       ia - IS-IS inter area, * - candidate default, U - per-user static route
       o - ODR, P - periodic downloaded static route

Gateway of last resort is not set

C    192.168.0.0/24 is directly connected, FastEthernet0/0
C    192.168.1.0/24 is directly connected, Serial0/1/0
```

图 2-3　路由器 R1 的路由表

步骤 5：在路由器 R2 的特权执行模式下输入 show ip route，查看路由表，如图 2-4 所示。

```
R2#show ip route
Codes: C - connected, S - static, R - RIP, M - mobile, B - BGP
       D - EIGRP, EX - EIGRP external, O - OSPF, IA - OSPF inter area
       N1 - OSPF NSSA external type 1, N2 - OSPF NSSA external type 2
       E1 - OSPF external type 1, E2 - OSPF external type 2
       i - IS-IS, su - IS-IS summary, L1 - IS-IS level-1, L2 - IS-IS level-2
       ia - IS-IS inter area, * - candidate default, U - per-user static route
       o - ODR, P - periodic downloaded static route

Gateway of last resort is not set

     172.16.0.0/24 is subnetted, 1 subnets
C       172.16.0.0 is directly connected, Serial0/1/1
C    192.168.1.0/24 is directly connected, Serial0/1/0
C    192.168.2.0/24 is directly connected, FastEthernet0/0
```

图 2-4　路由器 R2 的路由表

步骤 6：在路由器 R3 的特权执行模式下输入 show ip route，查看路由表，如图 2-5 所示。

```
R3#show ip route
Codes: C - connected, S - static, R - RIP, M - mobile, B - BGP
       D - EIGRP, EX - EIGRP external, O - OSPF, IA - OSPF inter area
       N1 - OSPF NSSA external type 1, N2 - OSPF NSSA external type 2
       E1 - OSPF external type 1, E2 - OSPF external type 2
       i - IS-IS, su - IS-IS summary, L1 - IS-IS level-1, L2 - IS-IS level-2
       ia - IS-IS inter area, * - candidate default, U - per-user static route
       o - ODR, P - periodic downloaded static route

Gateway of last resort is not set

     172.16.0.0/24 is subnetted, 2 subnets
C       172.16.0.0 is directly connected, Serial0/1/0
C       172.16.1.0 is directly connected, FastEthernet0/0
```

图 2-5　路由器 R3 的路由表

通过查看路由器 R1、R2 和 R3 的路由表可以发现，路由器 R1、R2 和 R3 对应的每个直连网络都有直连路由。直连路由没有问题。

步骤 7：在路由器 R1 的全局配置模式下配置直连静态路由，输入以下代码。

```
R1(config)#ip route 192.168.2.0 255.255.255.0 s0/1/0
R1(config)#ip route 172.16.0.0 255.255.255.0 s0/1/0
R1(config)#ip route 172.16.1.0 255.255.255.0 s0/1/0
```

步骤 8：在路由器 R2 的全局配置模式下配置直连静态路由，输入以下代码。

```
R2(config)#ip route 192.168.0.0 255.255.255.0 s0/1/0
R2(config)#ip route 172.16.1.0 255.255.255.0 s0/1/1
```

步骤 9：在路由器 R3 的全局配置模式下配置直连静态路由，输入以下代码。

```
R3(config)#ip route 192.168.0.0 255.255.255.0 s0/1/0
R3(config)#ip route 192.168.1.0 255.255.255.0 s0/1/0
R3(config)#ip route 192.168.2.0 255.255.255.0 s0/1/0
```

步骤 10：在路由器 R1 的特权执行模式下输入 show ip route，查看路由表，如图 2-6 所示。

```
R1#show ip route
Codes: C - connected, S - static, R - RIP, M - mobile, B - BGP
       D - EIGRP, EX - EIGRP external, O - OSPF, IA - OSPF inter area
       N1 - OSPF NSSA external type 1, N2 - OSPF NSSA external type 2
       E1 - OSPF external type 1, E2 - OSPF external type 2
       i - IS-IS, su - IS-IS summary, L1 - IS-IS level-1, L2 - IS-IS level-2
       ia - IS-IS inter area, * - candidate default, U - per-user static route
       o - ODR, P - periodic downloaded static route

Gateway of last resort is not set

     172.16.0.0/24 is subnetted, 2 subnets
S       172.16.0.0 is directly connected, Serial0/1/0
S       172.16.1.0 is directly connected, Serial0/1/0
C    192.168.0.0/24 is directly connected, FastEthernet0/0
C    192.168.1.0/24 is directly connected, Serial0/1/0
S    192.168.2.0/24 is directly connected, Serial0/1/0
```

图 2-6　路由器 R1 的路由表

步骤 11：在路由器 R2 的特权执行模式下输入 show ip route，查看路由表，如图 2-7 所示。

```
R2#show ip route
Codes: C - connected, S - static, R - RIP, M - mobile, B - BGP
       D - EIGRP, EX - EIGRP external, O - OSPF, IA - OSPF inter area
       N1 - OSPF NSSA external type 1, N2 - OSPF NSSA external type 2
       E1 - OSPF external type 1, E2 - OSPF external type 2
       i - IS-IS, su - IS-IS summary, L1 - IS-IS level-1, L2 - IS-IS level-2
       ia - IS-IS inter area, * - candidate default, U - per-user static route
       o - ODR, P - periodic downloaded static route

Gateway of last resort is not set

     172.16.0.0/24 is subnetted, 2 subnets
C       172.16.0.0 is directly connected, Serial0/1/1
S       172.16.1.0 is directly connected, Serial0/1/1
S    192.168.0.0/24 is directly connected, Serial0/1/0
C    192.168.1.0/24 is directly connected, Serial0/1/0
C    192.168.2.0/24 is directly connected, FastEthernet0/0
```

图 2-7　路由器 R2 的路由表

步骤 12：在路由器 R3 的特权执行模式下输入 show ip route，查看路由表，如图 2-8 所示。

```
R3#show ip route
Codes: C - connected, S - static, R - RIP, M - mobile, B - BGP
       D - EIGRP, EX - EIGRP external, O - OSPF, IA - OSPF inter area
       N1 - OSPF NSSA external type 1, N2 - OSPF NSSA external type 2
       E1 - OSPF external type 1, E2 - OSPF external type 2
       i - IS-IS, su - IS-IS summary, L1 - IS-IS level-1, L2 - IS-IS level-2
       ia - IS-IS inter area, * - candidate default, U - per-user static route
       o - ODR, P - periodic downloaded static route

Gateway of last resort is not set

     172.16.0.0/24 is subnetted, 2 subnets
C       172.16.0.0 is directly connected, Serial0/1/0
C       172.16.1.0 is directly connected, FastEthernet0/0
S    192.168.0.0/24 is directly connected, Serial0/1/0
S    192.168.1.0/24 is directly connected, Serial0/1/0
S    192.168.2.0/24 is directly connected, Serial0/1/0
```

图 2-8　路由器 R3 的路由表

步骤 13：在计算机 PC1 的命令行界面输入 ping 192.168.2.1，检验联通性，如图 2-9 所示。

```
C:\>ping 192.168.2.1

正在 Ping 192.168.2.1 具有 32 字节的数据：
来自 192.168.2.1 的回复：字节=32 时间=9ms TTL=126
来自 192.168.2.1 的回复：字节=32 时间=9ms TTL=126
来自 192.168.2.1 的回复：字节=32 时间=9ms TTL=126
来自 192.168.2.1 的回复：字节=32 时间=9ms TTL=126

192.168.2.1 的 Ping 统计信息：
    数据包：已发送 = 4，已接收 = 4，丢失 = 0 (0% 丢失)，
往返行程的估计时间(以毫秒为单位)：
    最短 = 9ms，最长 = 9ms，平均 = 9ms
```

图 2-9　从计算机 PC1 ping 计算机 PC2

步骤 14：在计算机 PC1 的命令行界面输入 ping 172.16.1.1，检验联通性，如图 2-10 所示。

```
C:\>ping 172.16.1.1

正在 Ping 172.16.1.1 具有 32 字节的数据：
来自 172.16.1.1 的回复：字节=32 时间=18ms TTL=125
来自 172.16.1.1 的回复：字节=32 时间=18ms TTL=125
来自 172.16.1.1 的回复：字节=32 时间=18ms TTL=125
来自 172.16.1.1 的回复：字节=32 时间=19ms TTL=125

172.16.1.1 的 Ping 统计信息：
    数据包：已发送 = 4，已接收 = 4，丢失 = 0 (0% 丢失)，
往返行程的估计时间(以毫秒为单位)：
    最短 = 18ms，最长 = 19ms，平均 = 18ms
```

图 2-10　从计算机 PC1 ping 计算机 PC3

步骤 15：在计算机 PC2 的命令行界面输入 ping 172.16.1.1，检验联通性，如图 2-11 所示。

```
C:\>ping 172.16.1.1

正在 Ping 172.16.1.1 具有 32 字节的数据：
来自 172.16.1.1 的回复：字节=32 时间=9ms TTL=126
来自 172.16.1.1 的回复：字节=32 时间=9ms TTL=126
来自 172.16.1.1 的回复：字节=32 时间=9ms TTL=126
来自 172.16.1.1 的回复：字节=32 时间=9ms TTL=126

172.16.1.1 的 Ping 统计信息：
    数据包：已发送 = 4，已接收 = 4，丢失 = 0 (0% 丢失)，
往返行程的估计时间(以毫秒为单位)：
    最短 = 9ms，最长 = 9ms，平均 = 9ms
```

图 2-11　从计算机 PC2 ping 计算机 PC3

通过查看路由器 R1、R2 和 R3 的路由表可以发现，直连静态路由在路由表中是路由来源为"S"的路由，3 台路由器 R1、R2 和 R3 的路由表中路由完整。通过从计算机 PC1 ping 计算机 PC2 和 PC3 可以发现，计算机 PC1、PC2 和 PC3 已经实现了互通。

2.4.2　下一跳静态路由的配置

本小节网络拓扑如图 2-1 所示，路由器 R1、R2 和 R3 的接口，以及计算机 PC1、PC2 和 PC3 的 IP 地址已经配置完成，要求完成下一跳静态路由的配置，实现计算机 PC1、PC2 和 PC3 互通。

步骤 1：在路由器 R1 的全局配置模式下配置下一跳静态路由，输入以下代码。

```
R1(config)#ip route 192.168.2.0 255.255.255.0 192.168.1.1
R1(config)#ip route 172.16.0.0 255.255.255.0 192.168.1.1
R1(config)#ip route 172.16.1.0 255.255.255.0 192.168.1.1
```

步骤 2：在路由器 R2 的全局配置模式下配置下一跳静态路由，输入以下代码。

```
R2(config)#ip route 192.168.0.0 255.255.255.0 192.168.1.2
R2(config)#ip route 172.16.1.0 255.255.255.0 172.16.0.2
```

步骤 3：在路由器 R3 的全局配置模式下配置下一跳静态路由，输入以下代码。

```
R3(config)#ip route 192.168.0.0 255.255.255.0 172.16.0.1
R3(config)#ip route 192.168.1.0 255.255.255.0 172.16.0.1
R3(config)#ip route 192.168.2.0 255.255.255.0 172.16.0.1
```

步骤 4：在路由器 R1 的特权执行模式下输入 show ip route，查看路由表，如图 2-12 所示。

```
R1#show ip route
Codes: C - connected, S - static, R - RIP, M - mobile, B - BGP
       D - EIGRP, EX - EIGRP external, O - OSPF, IA - OSPF inter area
       N1 - OSPF NSSA external type 1, N2 - OSPF NSSA external type 2
       E1 - OSPF external type 1, E2 - OSPF external type 2
       i - IS-IS, su - IS-IS summary, L1 - IS-IS level-1, L2 - IS-IS level-2
       ia - IS-IS inter area, * - candidate default, U - per-user static route
       o - ODR, P - periodic downloaded static route

Gateway of last resort is not set

     172.16.0.0/24 is subnetted, 2 subnets
S       172.16.0.0 [1/0] via 192.168.1.1
S       172.16.1.0 [1/0] via 192.168.1.1
C    192.168.0.0/24 is directly connected, FastEthernet0/0
C    192.168.1.0/24 is directly connected, Serial0/1/0
S    192.168.2.0/24 [1/0] via 192.168.1.1
```

图 2-12 路由器 R1 的路由表

步骤 5：在路由器 R2 的特权执行模式下输入 show ip route，查看路由表，如图 2-13 所示。

```
R2#show ip route
Codes: C - connected, S - static, R - RIP, M - mobile, B - BGP
       D - EIGRP, EX - EIGRP external, O - OSPF, IA - OSPF inter area
       N1 - OSPF NSSA external type 1, N2 - OSPF NSSA external type 2
       E1 - OSPF external type 1, E2 - OSPF external type 2
       i - IS-IS, su - IS-IS summary, L1 - IS-IS level-1, L2 - IS-IS level-2
       ia - IS-IS inter area, * - candidate default, U - per-user static route
       o - ODR, P - periodic downloaded static route

Gateway of last resort is not set

     172.16.0.0/24 is subnetted, 2 subnets
C       172.16.0.0 is directly connected, Serial0/1/1
S       172.16.1.0 [1/0] via 172.16.0.2
S    192.168.0.0/24 [1/0] via 192.168.1.2
C    192.168.1.0/24 is directly connected, Serial0/1/0
C    192.168.2.0/24 is directly connected, FastEthernet0/0
```

图 2-13 路由器 R2 的路由表

步骤 6：在路由器 R3 的特权执行模式下输入 show ip route，查看路由表，如图 2-14 所示。

```
R3#show ip route
Codes: C - connected, S - static, R - RIP, M - mobile, B - BGP
       D - EIGRP, EX - EIGRP external, O - OSPF, IA - OSPF inter area
       N1 - OSPF NSSA external type 1, N2 - OSPF NSSA external type 2
       E1 - OSPF external type 1, E2 - OSPF external type 2
       i - IS-IS, su - IS-IS summary, L1 - IS-IS level-1, L2 - IS-IS level-2
       ia - IS-IS inter area, * - candidate default, U - per-user static route
       o - ODR, P - periodic downloaded static route

Gateway of last resort is not set

     172.16.0.0/24 is subnetted, 2 subnets
C       172.16.0.0 is directly connected, Serial0/1/0
C       172.16.1.0 is directly connected, FastEthernet0/0
S    192.168.0.0/24 [1/0] via 172.16.0.1
S    192.168.1.0/24 [1/0] via 172.16.0.1
S    192.168.2.0/24 [1/0] via 172.16.0.1
```

图 2-14 路由器 R3 的路由表

步骤 7：在计算机 PC1 的命令行界面输入 ping 192.168.2.1，检验联通性，如图 2-15 所示。

```
C:\>ping 192.168.2.1

正在 Ping 192.168.2.1 具有 32 字节的数据：
来自 192.168.2.1 的回复: 字节=32 时间=9ms TTL=126
来自 192.168.2.1 的回复: 字节=32 时间=9ms TTL=126
来自 192.168.2.1 的回复: 字节=32 时间=9ms TTL=126
来自 192.168.2.1 的回复: 字节=32 时间=9ms TTL=126

192.168.2.1 的 Ping 统计信息:
    数据包: 已发送 = 4，已接收 = 4，丢失 = 0 (0% 丢失)，
往返行程的估计时间(以毫秒为单位):
    最短 = 9ms，最长 = 9ms，平均 = 9ms
```

图 2-15　从计算机 PC1 ping 计算机 PC2

步骤 8：在计算机 PC1 的命令行界面输入 ping 172.16.1.1，检验联通性，如图 2-16 所示。

```
C:\>ping 172.16.1.1

正在 Ping 172.16.1.1 具有 32 字节的数据：
来自 172.16.1.1 的回复: 字节=32 时间=18ms TTL=125
来自 172.16.1.1 的回复: 字节=32 时间=18ms TTL=125
来自 172.16.1.1 的回复: 字节=32 时间=18ms TTL=125
来自 172.16.1.1 的回复: 字节=32 时间=18ms TTL=125

172.16.1.1 的 Ping 统计信息:
    数据包: 已发送 = 4，已接收 = 4，丢失 = 0 (0% 丢失)，
往返行程的估计时间(以毫秒为单位):
    最短 = 18ms，最长 = 18ms，平均 = 18ms
```

图 2-16　从计算机 PC1 ping 计算机 PC3

步骤 9：在计算机 PC2 的命令行界面输入 ping 172.16.1.1，检验联通性，如图 2-17 所示。

```
C:\>ping 172.16.1.1

正在 Ping 172.16.1.1 具有 32 字节的数据：
来自 172.16.1.1 的回复: 字节=32 时间=9ms TTL=126
来自 172.16.1.1 的回复: 字节=32 时间=9ms TTL=126
来自 172.16.1.1 的回复: 字节=32 时间=9ms TTL=126
来自 172.16.1.1 的回复: 字节=32 时间=9ms TTL=126

172.16.1.1 的 Ping 统计信息:
    数据包: 已发送 = 4，已接收 = 4，丢失 = 0 (0% 丢失)，
往返行程的估计时间(以毫秒为单位):
    最短 = 9ms，最长 = 9ms，平均 = 9ms
```

图 2-17　从计算机 PC2 ping 计算机 PC3

通过查看 3 台路由器 R1、R2 和 R3 的路由表与检验联通性可以发现，3 台路由器 R1、R2 和 R3 的路由表中路由完整，计算机 PC1、PC2 和 PC3 实现了互通。

2.4.3　完全指定静态路由的配置

本小节网络拓扑如图 2-1 所示，路由器 R1、R2 和 R3 的接口，以及计算机 PC1、PC2 和 PC3 的 IP 地址已经配置完成，要求完成完全指定静态路由的配置，实现计算机 PC1、PC2 和 PC3 互通。

步骤 1：在路由器 R1 的全局配置模式下配置完全指定静态路由，输入以下代码。

```
R1(config)#ip route 192.168.2.0 255.255.255.0 s0/1/0 192.168.1.1
R1(config)#ip route 172.16.0.0 255.255.255.0 s0/1/0 192.168.1.1
R1(config)#ip route 172.16.1.0 255.255.255.0 s0/1/0 192.168.1.1
```

步骤 2：在路由器 R2 的全局配置模式下配置完全指定静态路由，输入以下代码。

```
R2(config)#ip route 192.168.0.0 255.255.255.0 s0/1/0 192.168.1.2
R2(config)#ip route 172.16.1.0 255.255.255.0 s0/1/1 172.16.0.2
```

步骤 3：在路由器 R3 的全局配置模式下配置完全指定静态路由，输入以下代码。

```
R3(config)#ip route 192.168.0.0 255.255.255.0 s0/1/0 172.16.0.1
R3(config)#ip route 192.168.1.0 255.255.255.0 s0/1/0 172.16.0.1
R3(config)#ip route 192.168.2.0 255.255.255.0 s0/1/0 172.16.0.1
```

步骤 4：在路由器 R1 的特权执行模式下输入 show ip route，查看路由表，如图 2-18 所示。

```
R1#show ip route
Codes: C - connected, S - static, R - RIP, M - mobile, B - BGP
       D - EIGRP, EX - EIGRP external, O - OSPF, IA - OSPF inter area
       N1 - OSPF NSSA external type 1, N2 - OSPF NSSA external type 2
       E1 - OSPF external type 1, E2 - OSPF external type 2
       i - IS-IS, su - IS-IS summary, L1 - IS-IS level-1, L2 - IS-IS level-2
       ia - IS-IS inter area, * - candidate default, U - per-user static route
       o - ODR, P - periodic downloaded static route

Gateway of last resort is not set

     172.16.0.0/24 is subnetted, 2 subnets
S       172.16.0.0 [1/0] via 192.168.1.1, Serial0/1/0
S       172.16.1.0 [1/0] via 192.168.1.1, Serial0/1/0
C    192.168.0.0/24 is directly connected, FastEthernet0/0
C    192.168.1.0/24 is directly connected, Serial0/1/0
S    192.168.2.0/24 [1/0] via 192.168.1.1, Serial0/1/0
```

图 2-18　路由器 R1 的路由表

步骤 5：在路由器 R2 的特权执行模式下输入 show ip route，查看路由表，如图 2-19 所示。

```
R2#show ip route
Codes: C - connected, S - static, R - RIP, M - mobile, B - BGP
       D - EIGRP, EX - EIGRP external, O - OSPF, IA - OSPF inter area
       N1 - OSPF NSSA external type 1, N2 - OSPF NSSA external type 2
       E1 - OSPF external type 1, E2 - OSPF external type 2
       i - IS-IS, su - IS-IS summary, L1 - IS-IS level-1, L2 - IS-IS level-2
       ia - IS-IS inter area, * - candidate default, U - per-user static route
       o - ODR, P - periodic downloaded static route

Gateway of last resort is not set

     172.16.0.0/24 is subnetted, 2 subnets
C       172.16.0.0 is directly connected, Serial0/1/1
S       172.16.1.0 [1/0] via 172.16.0.2, Serial0/1/1
S    192.168.0.0/24 [1/0] via 192.168.1.2, Serial0/1/0
C    192.168.1.0/24 is directly connected, Serial0/1/0
C    192.168.2.0/24 is directly connected, FastEthernet0/0
```

图 2-19　路由器 R2 的路由表

步骤 6：在路由器 R3 的特权执行模式下输入 show ip route，查看路由表，如图 2-20 所示。

```
R3#show ip route
Codes: C - connected, S - static, R - RIP, M - mobile, B - BGP
       D - EIGRP, EX - EIGRP external, O - OSPF, IA - OSPF inter area
       N1 - OSPF NSSA external type 1, N2 - OSPF NSSA external type 2
       E1 - OSPF external type 1, E2 - OSPF external type 2
       i - IS-IS, su - IS-IS summary, L1 - IS-IS level-1, L2 - IS-IS level-2
       ia - IS-IS inter area, * - candidate default, U - per-user static route
       o - ODR, P - periodic downloaded static route

Gateway of last resort is not set

     172.16.0.0/24 is subnetted, 2 subnets
C       172.16.0.0 is directly connected, Serial0/1/0
C       172.16.1.0 is directly connected, FastEthernet0/0
S    192.168.0.0/24 [1/0] via 172.16.0.1, Serial0/1/0
S    192.168.1.0/24 [1/0] via 172.16.0.1, Serial0/1/0
S    192.168.2.0/24 [1/0] via 172.16.0.1, Serial0/1/0
```

图 2-20　路由器 R3 的路由表

步骤 7：在计算机 PC1 的命令行界面输入 ping 192.168.2.1，检验联通性，如图 2-21 所示。

```
C:\>ping 192.168.2.1

正在 Ping 192.168.2.1 具有 32 字节的数据:
来自 192.168.2.1 的回复: 字节=32 时间=9ms TTL=126
来自 192.168.2.1 的回复: 字节=32 时间=9ms TTL=126
来自 192.168.2.1 的回复: 字节=32 时间=9ms TTL=126
来自 192.168.2.1 的回复: 字节=32 时间=9ms TTL=126

192.168.2.1 的 Ping 统计信息:
    数据包: 已发送 = 4，已接收 = 4，丢失 = 0 (0% 丢失)，
往返行程的估计时间(以毫秒为单位):
    最短 = 9ms，最长 = 9ms，平均 = 9ms
```

图 2-21　从计算机 PC1 ping 计算机 PC2

步骤 8：在计算机 PC1 的命令行界面输入 ping 172.16.1.1，检验联通性，如图 2-22 所示。

```
C:\>ping 172.16.1.1

正在 Ping 172.16.1.1 具有 32 字节的数据:
来自 172.16.1.1 的回复: 字节=32 时间=19ms TTL=125
来自 172.16.1.1 的回复: 字节=32 时间=19ms TTL=125
来自 172.16.1.1 的回复: 字节=32 时间=19ms TTL=125
来自 172.16.1.1 的回复: 字节=32 时间=18ms TTL=125

172.16.1.1 的 Ping 统计信息:
    数据包: 已发送 = 4，已接收 = 4，丢失 = 0 (0% 丢失)，
往返行程的估计时间(以毫秒为单位):
    最短 = 18ms，最长 = 19ms，平均 = 18ms
```

图 2-22　从计算机 PC1 ping 计算机 PC3

步骤 9：在计算机 PC2 的命令行界面输入 ping 172.16.1.1，检验联通性，如图 2-23 所示。

```
C:\>ping 172.16.1.1

正在 Ping 172.16.1.1 具有 32 字节的数据:
来自 172.16.1.1 的回复: 字节=32 时间=9ms TTL=126
来自 172.16.1.1 的回复: 字节=32 时间=9ms TTL=126
来自 172.16.1.1 的回复: 字节=32 时间=9ms TTL=126
来自 172.16.1.1 的回复: 字节=32 时间=9ms TTL=126

172.16.1.1 的 Ping 统计信息:
    数据包: 已发送 = 4，已接收 = 4，丢失 = 0 (0% 丢失)，
往返行程的估计时间(以毫秒为单位):
    最短 = 9ms，最长 = 9ms，平均 = 9ms
```

图 2-23　从计算机 PC2 ping 计算机 PC3

通过查看 3 台路由器 R1、R2 和 R3 的路由表与检验联通性可以发现，3 台路由器 R1、R2 和 R3 的路由表中路由完整，计算机 PC1、PC2 和 PC3 实现了互通。

2.4.4　默认路由的配置

本小节网络拓扑如图 2-1 所示，路由器 R1、R2 和 R3 的接口，以及计算机 PC1、PC2 和 PC3 的 IP 地址已经配置完成，要求为路由器 R1 和 R2 配置标准静态路由，为路由器 R3 配置默认路由，实现计算机 PC1、PC2 和 PC3 互通。

步骤 1：在路由器 R1 的全局配置模式下配置标准静态路由，输入以下代码。

```
R1(config)#ip route 192.168.2.0 255.255.255.0 192.168.1.1
R1(config)#ip route 172.16.0.0 255.255.255.0 192.168.1.1
```

```
R1(config)#ip route 172.16.1.0 255.255.255.0 192.168.1.1
```

步骤 2：在路由器 R2 的全局配置模式下配置标准静态路由，输入以下代码。

```
R2(config)#ip route 192.168.0.0 255.255.255.0 192.168.1.2
R2(config)#ip route 172.16.1.0 255.255.255.0 172.16.0.2
```

步骤 3：在路由器 R3 的全局配置模式下配置默认路由，输入以下代码。

```
R3(config)#ip route 0.0.0.0 0.0.0.0 172.16.0.1
```

步骤 4：在路由器 R1 的特权执行模式下输入 show ip route，查看路由表，如图 2-24 所示。

```
R1#show ip route
Codes: C - connected, S - static, R - RIP, M - mobile, B - BGP
       D - EIGRP, EX - EIGRP external, O - OSPF, IA - OSPF inter area
       N1 - OSPF NSSA external type 1, N2 - OSPF NSSA external type 2
       E1 - OSPF external type 1, E2 - OSPF external type 2
       i - IS-IS, su - IS-IS summary, L1 - IS-IS level-1, L2 - IS-IS level-2
       ia - IS-IS inter area, * - candidate default, U - per-user static route
       o - ODR, P - periodic downloaded static route

Gateway of last resort is not set

     172.16.0.0/24 is subnetted, 2 subnets
S       172.16.0.0 [1/0] via 192.168.1.1
S       172.16.1.0 [1/0] via 192.168.1.1
C    192.168.0.0/24 is directly connected, FastEthernet0/0
C    192.168.1.0/24 is directly connected, Serial0/1/0
S    192.168.2.0/24 [1/0] via 192.168.1.1
```

图 2-24　路由器 R1 的路由表

步骤 5：在路由器 R2 的特权执行模式下输入 show ip route，查看路由表，如图 2-25 所示。

```
R2#show ip route
Codes: C - connected, S - static, R - RIP, M - mobile, B - BGP
       D - EIGRP, EX - EIGRP external, O - OSPF, IA - OSPF inter area
       N1 - OSPF NSSA external type 1, N2 - OSPF NSSA external type 2
       E1 - OSPF external type 1, E2 - OSPF external type 2
       i - IS-IS, su - IS-IS summary, L1 - IS-IS level-1, L2 - IS-IS level-2
       ia - IS-IS inter area, * - candidate default, U - per-user static route
       o - ODR, P - periodic downloaded static route

Gateway of last resort is not set

     172.16.0.0/24 is subnetted, 2 subnets
C       172.16.0.0 is directly connected, Serial0/1/1
S       172.16.1.0 [1/0] via 172.16.0.2
S    192.168.0.0/24 [1/0] via 192.168.1.2
C    192.168.1.0/24 is directly connected, Serial0/1/0
C    192.168.2.0/24 is directly connected, FastEthernet0/0
```

图 2-25　路由器 R2 的路由表

步骤 6：在路由器 R3 的特权执行模式下输入 show ip route，查看路由表，如图 2-26 所示。

```
R3#show ip route
Codes: C - connected, S - static, R - RIP, M - mobile, B - BGP
       D - EIGRP, EX - EIGRP external, O - OSPF, IA - OSPF inter area
       N1 - OSPF NSSA external type 1, N2 - OSPF NSSA external type 2
       E1 - OSPF external type 1, E2 - OSPF external type 2
       i - IS-IS, su - IS-IS summary, L1 - IS-IS level-1, L2 - IS-IS level-2
       ia - IS-IS inter area, * - candidate default, U - per-user static route
       o - ODR, P - periodic downloaded static route

Gateway of last resort is 172.16.0.1 to network 0.0.0.0

     172.16.0.0/24 is subnetted, 2 subnets
C       172.16.0.0 is directly connected, Serial0/1/0
C       172.16.1.0 is directly connected, FastEthernet0/0
S*   0.0.0.0/0 [1/0] via 172.16.0.1
```

图 2-26　路由器 R3 的路由表

步骤 7：在计算机 PC1 的命令行界面输入 ping 192.168.2.1，检验联通性，如图 2-27 所示。

```
C:\>ping 192.168.2.1
正在 Ping 192.168.2.1 具有 32 字节的数据：
来自 192.168.2.1 的回复：字节=32 时间=9ms TTL=126
来自 192.168.2.1 的回复：字节=32 时间=9ms TTL=126
来自 192.168.2.1 的回复：字节=32 时间=9ms TTL=126
来自 192.168.2.1 的回复：字节=32 时间=9ms TTL=126

192.168.2.1 的 Ping 统计信息：
    数据包：已发送 = 4，已接收 = 4，丢失 = 0 (0% 丢失)，
往返行程的估计时间(以毫秒为单位)：
    最短 = 9ms，最长 = 9ms，平均 = 9ms
```

图 2-27　从计算机 PC1 ping 计算机 PC2

步骤 8：在计算机 PC1 的命令行界面输入 ping 172.16.1.1，检验联通性，如图 2-28 所示。

```
C:\>ping 172.16.1.1
正在 Ping 172.16.1.1 具有 32 字节的数据：
来自 172.16.1.1 的回复：字节=32 时间=18ms TTL=125
来自 172.16.1.1 的回复：字节=32 时间=19ms TTL=125
来自 172.16.1.1 的回复：字节=32 时间=19ms TTL=125
来自 172.16.1.1 的回复：字节=32 时间=18ms TTL=125

172.16.1.1 的 Ping 统计信息：
    数据包：已发送 = 4，已接收 = 4，丢失 = 0 (0% 丢失)，
往返行程的估计时间(以毫秒为单位)：
    最短 = 18ms，最长 = 19ms，平均 = 18ms
```

图 2-28　从计算机 PC1 ping 计算机 PC3

步骤 9：在计算机 PC2 的命令行界面输入 ping 172.16.1.1，检验联通性，如图 2-29 所示。

```
C:\>ping 172.16.1.1
正在 Ping 172.16.1.1 具有 32 字节的数据：
来自 172.16.1.1 的回复：字节=32 时间=9ms TTL=126
来自 172.16.1.1 的回复：字节=32 时间=9ms TTL=126
来自 172.16.1.1 的回复：字节=32 时间=9ms TTL=126
来自 172.16.1.1 的回复：字节=32 时间=9ms TTL=126

172.16.1.1 的 Ping 统计信息：
    数据包：已发送 = 4，已接收 = 4，丢失 = 0 (0% 丢失)，
往返行程的估计时间(以毫秒为单位)：
    最短 = 9ms，最长 = 9ms，平均 = 9ms
```

图 2-29　从计算机 PC2 ping 计算机 PC3

通过查看路由器 R1、R2 和 R3 这 3 台路由器的路由表与检验联通性可以发现，默认路由在路由表中是路由来源为"S*"的路由。3 台路由器 R1、R2 和 R3 的路由表中路由完整，计算机 PC1、PC2 和 PC3 实现了互通。

2.4.5　汇总静态路由的配置

本小节网络拓扑如图 2-1 所示，路由器 R1、R2 和 R3 的接口，以及计算机 PC1、PC2 和 PC3 的 IP 地址已经配置完成，要求为路由器 R1 和 R2 配置标准静态路由，为路由器 R1 和 R3 配置汇总静态路由，实现计算机 PC1、PC2 和 PC3 互通。

步骤 1：在路由器 R1 的全局配置模式下配置标准静态路由和汇总静态路由，输入以下代码。

```
R1(config)#ip route 192.168.2.0 255.255.255.0 192.168.1.1
R1(config)#ip route 172.16.0.0 255.255.254.0 192.168.1.1
```

步骤 2：在路由器 R2 的全局配置模式下配置标准静态路由，输入以下代码。

```
R2(config)#ip route 192.168.0.0 255.255.255.0 192.168.1.2
R2(config)#ip route 172.16.1.0 255.255.255.0 172.16.0.2
```

步骤 3：在路由器 R3 的全局配置模式下配置汇总静态路由，输入以下代码。

```
R3(config)#ip route 192.168.0.0 255.255.252.0 172.16.0.1
```

本小节中，路由器 R3 把数据包发往 192.168.0.0/24、192.168.1.0/24 和 192.168.2.0/24 这 3 个网络，路由器 R3 的下一跳 IP 地址是相同的，此时可以把这 3 个网络地址汇总成 192.168.0.0/22，然后在配置静态路由时使用汇总后的网络地址，这样配置的就是汇总静态路由。

步骤 4：在路由器 R1 的特权执行模式下输入 show ip route，查看路由表，如图 2-30 所示。

```
R1#show ip route
Codes: C - connected, S - static, R - RIP, M - mobile, B - BGP
       D - EIGRP, EX - EIGRP external, O - OSPF, IA - OSPF inter area
       N1 - OSPF NSSA external type 1, N2 - OSPF NSSA external type 2
       E1 - OSPF external type 1, E2 - OSPF external type 2
       i - IS-IS, su - IS-IS summary, L1 - IS-IS level-1, L2 - IS-IS level-2
       ia - IS-IS inter area, * - candidate default, U - per-user static route
       o - ODR, P - periodic downloaded static route

Gateway of last resort is not set

     172.16.0.0/23 is subnetted, 1 subnets
S       172.16.0.0 [1/0] via 192.168.1.1
C    192.168.0.0/24 is directly connected, FastEthernet0/0
C    192.168.1.0/24 is directly connected, Serial0/1/0
S    192.168.2.0/24 [1/0] via 192.168.1.1
```

图 2-30　路由器 R1 的路由表

步骤 5：在路由器 R2 的特权执行模式下输入 show ip route，查看路由表，如图 2-31 所示。

```
R2#show ip route
Codes: C - connected, S - static, R - RIP, M - mobile, B - BGP
       D - EIGRP, EX - EIGRP external, O - OSPF, IA - OSPF inter area
       N1 - OSPF NSSA external type 1, N2 - OSPF NSSA external type 2
       E1 - OSPF external type 1, E2 - OSPF external type 2
       i - IS-IS, su - IS-IS summary, L1 - IS-IS level-1, L2 - IS-IS level-2
       ia - IS-IS inter area, * - candidate default, U - per-user static route
       o - ODR, P - periodic downloaded static route

Gateway of last resort is not set

     172.16.0.0/24 is subnetted, 2 subnets
C       172.16.0.0 is directly connected, Serial0/1/1
S       172.16.1.0 [1/0] via 172.16.0.2
S    192.168.0.0/24 [1/0] via 192.168.1.2
C    192.168.1.0/24 is directly connected, Serial0/1/0
C    192.168.2.0/24 is directly connected, FastEthernet0/0
```

图 2-31　路由器 R2 的路由表

步骤 6：在路由器 R3 的特权执行模式下输入 show ip route，查看路由表，如图 2-32 所示。

```
R3#show ip route
Codes: C - connected, S - static, R - RIP, M - mobile, B - BGP
       D - EIGRP, EX - EIGRP external, O - OSPF, IA - OSPF inter area
       N1 - OSPF NSSA external type 1, N2 - OSPF NSSA external type 2
       E1 - OSPF external type 1, E2 - OSPF external type 2
       i - IS-IS, su - IS-IS summary, L1 - IS-IS level-1, L2 - IS-IS level-2
       ia - IS-IS inter area, * - candidate default, U - per-user static route
       o - ODR, P - periodic downloaded static route

Gateway of last resort is not set

     172.16.0.0/24 is subnetted, 2 subnets
C       172.16.0.0 is directly connected, Serial0/1/0
C       172.16.1.0 is directly connected, FastEthernet0/0
S    192.168.0.0/22 [1/0] via 172.16.0.1
```

图 2-32　路由器 R3 的路由表

步骤 7：在计算机 PC1 的命令行界面输入 ping 192.168.2.1，检验联通性，如图 2-33 所示。

```
C:\>ping 192.168.2.1

正在 Ping 192.168.2.1 具有 32 字节的数据：
来自 192.168.2.1 的回复：字节=32 时间=9ms TTL=126
来自 192.168.2.1 的回复：字节=32 时间=9ms TTL=126
来自 192.168.2.1 的回复：字节=32 时间=9ms TTL=126
来自 192.168.2.1 的回复：字节=32 时间=9ms TTL=126

192.168.2.1 的 Ping 统计信息：
    数据包：已发送 = 4，已接收 = 4，丢失 = 0 (0% 丢失)，
往返行程的估计时间(以毫秒为单位)：
    最短 = 9ms，最长 = 9ms，平均 = 9ms
```

图 2-33　从计算机 PC1 ping 计算机 PC2

步骤 8：在计算机 PC1 的命令行界面输入 ping 172.16.1.1，检验联通性，如图 2-34 所示。

```
C:\>ping 172.16.1.1

正在 Ping 172.16.1.1 具有 32 字节的数据：
来自 172.16.1.1 的回复：字节=32 时间=18ms TTL=125
来自 172.16.1.1 的回复：字节=32 时间=19ms TTL=125
来自 172.16.1.1 的回复：字节=32 时间=18ms TTL=125
来自 172.16.1.1 的回复：字节=32 时间=18ms TTL=125

172.16.1.1 的 Ping 统计信息：
    数据包：已发送 = 4，已接收 = 4，丢失 = 0 (0% 丢失)，
往返行程的估计时间(以毫秒为单位)：
    最短 = 18ms，最长 = 19ms，平均 = 18ms
```

图 2-34　从计算机 PC1 ping 计算机 PC3

步骤 9：在计算机 PC2 的命令行界面输入 ping 172.16.1.1，检验联通性，如图 2-35 所示。

```
C:\>ping 172.16.1.1

正在 Ping 172.16.1.1 具有 32 字节的数据：
来自 172.16.1.1 的回复：字节=32 时间=9ms TTL=126
来自 172.16.1.1 的回复：字节=32 时间=9ms TTL=126
来自 172.16.1.1 的回复：字节=32 时间=9ms TTL=126
来自 172.16.1.1 的回复：字节=32 时间=9ms TTL=126

172.16.1.1 的 Ping 统计信息：
    数据包：已发送 = 4，已接收 = 4，丢失 = 0 (0% 丢失)，
往返行程的估计时间(以毫秒为单位)：
    最短 = 9ms，最长 = 9ms，平均 = 9ms
```

图 2-35　从计算机 PC2 ping 计算机 PC3

通过查看路由表可以发现，使用汇总后的网络地址，把原来配置的 3 条标准静态路由用一条汇总静态路由代替，路由表的长度缩短，从而加快路由查找过程。通过从计算机 PC1 ping 计算机 PC2 和 PC3 可以发现，计算机 PC1、PC2 和 PC3 实现了互通。

2.4.6　浮动静态路由的配置

某学校网络拓扑如图 2-36 所示，由主校区网络和分校区网络组成，主校区和分校区的网络采用双链路互联，R1-R2 为主路径，R1-R3 为备用路径，路由器 R1、R2、R3 和 R4 的接口已经配置完成，要求：为路由器 R2、R3 和 R4 配置标准静态路由，为路由器 R1 配置浮动静态路由，实现主校区与分校区互联的主路径和备用路径的自动切换。

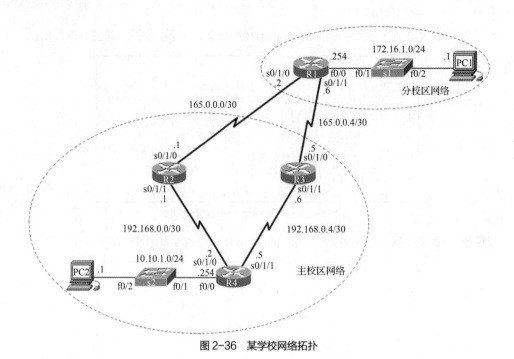

图 2-36 某学校网络拓扑

步骤 1：在路由器 R2 的全局配置模式下配置标准静态路由，输入以下代码。

```
R2(config)#ip route 10.10.1.0 255.255.255.0 192.168.0.2
R2(config)#ip route 172.16.1.0 255.255.255.0 165.0.0.2
```

步骤 2：在路由器 R3 的全局配置模式下配置标准静态路由，输入以下代码。

```
R3(config)#ip route 10.10.1.0 255.255.255.0 192.168.0.5
R3(config)#ip route 172.16.1.0 255.255.255.0 165.0.0.6
```

步骤 3：在路由器 R4 的全局配置模式下配置标准静态路由，输入以下代码。

```
R4(config)#ip route 172.16.1.0 255.255.255.0 192.168.0.6
```

步骤 4：在路由器 R1 的全局配置模式下配置浮动静态路由，输入以下代码。

```
R1(config)#ip route 0.0.0.0 0.0.0.0 165.0.0.1
R1(config)#ip route 0.0.0.0 0.0.0.0 165.0.0.5 5
```

步骤 5：在路由器 R1 的特权执行模式下输入 show ip route，查看路由表，如图 2-37 所示。

```
R1#show ip route
Codes: C - connected, S - static, R - RIP, M - mobile, B - BGP
       D - EIGRP, EX - EIGRP external, O - OSPF, IA - OSPF inter area
       N1 - OSPF NSSA external type 1, N2 - OSPF NSSA external type 2
       E1 - OSPF external type 1, E2 - OSPF external type 2
       i - IS-IS, su - IS-IS summary, L1 - IS-IS level-1, L2 - IS-IS level-2
       ia - IS-IS inter area, * - candidate default, U - per-user static route
       o - ODR, P - periodic downloaded static route

Gateway of last resort is 165.0.0.1 to network 0.0.0.0

     172.16.0.0/24 is subnetted, 1 subnets
C       172.16.1.0 is directly connected, FastEthernet0/0
     165.0.0.0/30 is subnetted, 2 subnets
C       165.0.0.4 is directly connected, Serial0/1/1
C       165.0.0.0 is directly connected, Serial0/1/0
S*   0.0.0.0/0 [1/0] via 165.0.0.1
```

图 2-37 路由器 R1 的路由表

步骤 6：在路由器 R2 的特权执行模式下输入 show ip route，查看路由表，如图 2-38 所示。

```
R2#show ip route
Codes: C - connected, S - static, R - RIP, M - mobile, B - BGP
       D - EIGRP, EX - EIGRP external, O - OSPF, IA - OSPF inter area
       N1 - OSPF NSSA external type 1, N2 - OSPF NSSA external type 2
       E1 - OSPF external type 1, E2 - OSPF external type 2
       i - IS-IS, su - IS-IS summary, L1 - IS-IS level-1, L2 - IS-IS level-2
       ia - IS-IS inter area, * - candidate default, U - per-user static route
       o - ODR, P - periodic downloaded static route

Gateway of last resort is not set

     172.16.0.0/24 is subnetted, 1 subnets
S       172.16.1.0 [1/0] via 165.0.0.2
     10.0.0.0/24 is subnetted, 1 subnets
S       10.10.1.0 [1/0] via 192.168.0.2
     192.168.0.0/30 is subnetted, 1 subnets
C       192.168.0.0 is directly connected, Serial0/1/1
     165.0.0.0/30 is subnetted, 1 subnets
C       165.0.0.0 is directly connected, Serial0/1/0
```

图 2-38　路由器 R2 的路由表

步骤 7：在路由器 R3 的特权执行模式下输入 show ip route，查看路由表，如图 2-39 所示。

```
R3#show ip route
Codes: C - connected, S - static, R - RIP, M - mobile, B - BGP
       D - EIGRP, EX - EIGRP external, O - OSPF, IA - OSPF inter area
       N1 - OSPF NSSA external type 1, N2 - OSPF NSSA external type 2
       E1 - OSPF external type 1, E2 - OSPF external type 2
       i - IS-IS, su - IS-IS summary, L1 - IS-IS level-1, L2 - IS-IS level-2
       ia - IS-IS inter area, * - candidate default, U - per-user static route
       o - ODR, P - periodic downloaded static route

Gateway of last resort is not set

     172.16.0.0/24 is subnetted, 1 subnets
S       172.16.1.0 [1/0] via 165.0.0.6
     10.0.0.0/24 is subnetted, 1 subnets
S       10.10.1.0 [1/0] via 192.168.0.5
C    192.168.0.0/24 is directly connected, Serial0/1/1
     165.0.0.0/30 is subnetted, 1 subnets
C       165.0.0.4 is directly connected, Serial0/1/0
```

图 2-39　路由器 R3 的路由表

步骤 8：在路由器 R4 的特权执行模式下输入 show ip route，查看路由表，如图 2-40 所示。

```
R4#show ip route
Codes: C - connected, S - static, R - RIP, M - mobile, B - BGP
       D - EIGRP, EX - EIGRP external, O - OSPF, IA - OSPF inter area
       N1 - OSPF NSSA external type 1, N2 - OSPF NSSA external type 2
       E1 - OSPF external type 1, E2 - OSPF external type 2
       i - IS-IS, su - IS-IS summary, L1 - IS-IS level-1, L2 - IS-IS level-2
       ia - IS-IS inter area, * - candidate default, U - per-user static route
       o - ODR, P - periodic downloaded static route

Gateway of last resort is not set

     172.16.0.0/24 is subnetted, 1 subnets
S       172.16.1.0 [1/0] via 192.168.0.6
     10.0.0.0/24 is subnetted, 1 subnets
C       10.10.1.0 is directly connected, FastEthernet0/0
     192.168.0.0/30 is subnetted, 2 subnets
C       192.168.0.0 is directly connected, Serial0/1/0
C       192.168.0.4 is directly connected, Serial0/1/1
```

图 2-40　路由器 R4 的路由表

步骤 9：在计算机 PC1 的命令行界面输入 ping 10.10.1.1，检验联通性，如图 2-41 所示。

```
C:\>ping 10.10.1.1

正在 Ping 10.10.1.1 具有 32 字节的数据：
来自 10.10.1.1 的回复: 字节=32 时间=28ms TTL=125
来自 10.10.1.1 的回复: 字节=32 时间=27ms TTL=125
来自 10.10.1.1 的回复: 字节=32 时间=28ms TTL=125
来自 10.10.1.1 的回复: 字节=32 时间=27ms TTL=125

10.10.1.1 的 Ping 统计信息：
    数据包: 已发送 = 4，已接收 = 4，丢失 = 0 (0% 丢失)，
往返行程的估计时间(以毫秒为单位)：
    最短 = 27ms，最长 = 28ms，平均 = 27ms
```

图 2-41　从计算机 PC1 ping 计算机 PC2

步骤 10：在计算机 PC1 的命令行界面输入 tracert 10.10.1.1，跟踪数据包从计算机 PCI 访问计算机 PC2 的路径，如图 2-42 所示。

```
C:\>tracert 10.10.1.1

通过最多 30 个跃点跟踪到 10.10.1.1 的路由

  1    <1 毫秒    <1 毫秒    <1 毫秒  172.16.1.254
  2    11 ms      11 ms      11 ms    165.0.0.1
  3    31 ms      31 ms      31 ms    192.168.0.2
  4    38 ms      38 ms      38 ms    10.10.1.1

跟踪完成。
```

图 2-42　跟踪数据包从计算机 PC1 访问计算机 PC2 的路径

步骤 11：在路由器 R1 的全局配置模式下关闭 s0/1/0 接口，输入以下代码。

```
R1(config)#interface s0/1/0
R1(config-if)#shutdown
```

步骤 12：在路由器 R1 的特权执行模式下输入 show ip route，查看路由表，如图 2-43 所示。

```
R1#show ip route
Codes: C - connected, S - static, R - RIP, M - mobile, B - BGP
       D - EIGRP, EX - EIGRP external, O - OSPF, IA - OSPF inter area
       N1 - OSPF NSSA external type 1, N2 - OSPF NSSA external type 2
       E1 - OSPF external type 1, E2 - OSPF external type 2
       i - IS-IS, su - IS-IS summary, L1 - IS-IS level-1, L2 - IS-IS level-2
       ia - IS-IS inter area, * - candidate default, U - per-user static route
       o - ODR, P - periodic downloaded static route

Gateway of last resort is 165.0.0.5 to network 0.0.0.0

     172.16.0.0/24 is subnetted, 1 subnets
C       172.16.1.0 is directly connected, FastEthernet0/0
     165.0.0.0/30 is subnetted, 1 subnets
C       165.0.0.4 is directly connected, Serial0/1/1
S*   0.0.0.0/0 [5/0] via 165.0.0.5
```

图 2-43　路由器 R1 的路由表

步骤 13：在计算机 PC1 的命令行界面输入 ping 10.10.1.1，检验联通性，如图 2-44 所示。

```
C:\>ping 10.10.1.1

正在 Ping 10.10.1.1 具有 32 字节的数据:
来自 10.10.1.1 的回复: 字节=32 时间=41ms TTL=125
来自 10.10.1.1 的回复: 字节=32 时间=43ms TTL=125
来自 10.10.1.1 的回复: 字节=32 时间=41ms TTL=125
来自 10.10.1.1 的回复: 字节=32 时间=42ms TTL=125

10.10.1.1 的 Ping 统计信息:
    数据包: 已发送 = 4，已接收 = 4，丢失 = 0 (0% 丢失)，
往返行程的估计时间(以毫秒为单位):
    最短 = 41ms，最长 = 43ms，平均 = 41ms
```

图 2-44　从计算机 PC1 ping 计算机 PC2

步骤 14：在计算机 PC1 的命令行界面输入 tracert 10.10.1.1，跟踪数据包从计算机 PCI 访问计算机 PC2 的路径，如图 2-45 所示。

```
C:\>tracert 10.10.1.1

通过最多 30 个跃点跟踪到 10.10.1.1 的路由

  1    <1 毫秒    <1 毫秒    <1 毫秒  172.16.1.254
  2    16 ms      17 ms      15 ms    165.0.0.5
  3    47 ms      47 ms      46 ms    192.168.0.5
  4    56 ms      54 ms      54 ms    10.10.1.1

跟踪完成。
```

图 2-45　跟踪数据包从计算机 PC1 访问计算机 PC2 的路径

通过查看路由表和 tracert 命令的执行结果可以发现，当主校区和分校区互联的链路都正常时，数据包从计算机 PC1 去往计算机 PC2，走的是 R1-R2-R4 这条链路；当路由器 R1 与 R2 互联的链路出现故障时（本小节把路由器 R1 互联路由器 R2 的 s0/1/0 接口禁用了），数据包从计算机 PC1 去往计算机 PC2 时会自动浮动到 R1-R3-R4。数据从计算机 PC2 返回计算机 PC1 时也可以通过为路由器 R4 配置浮动静态路由，使 R4-R3-R1 为主路径、R4-R2-R1 为备用路径。

2.5 项目小结

本项目完成了直连静态路由、下一跳静态路由、完全指定静态路由、默认路由、汇总静态路由和浮动静态路由的配置，静态路由是管理员用 ip route 命令手动配置的将数据包转发到目的网络的路径。直连路由是静态路由和动态路由生成的基础，所以在配置静态路由前应先检查直连路由是否完整。如果路由器的路由表中没有直连路由，那么静态路由和动态路由在路由器的路由表中都不会生成。

2.6 拓展训练

本项目的拓展训练网络拓扑如图 2-46 所示，要求完成如下配置：
（1）完成路由器接口和计算机 IP 地址的配置；
（2）完成路由器 R1 的汇总静态路由的配置；
（3）完成路由器 R2 的默认路由的配置，实现网络互通。

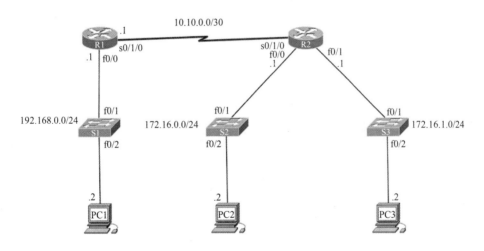

图 2-46　项目 2 拓展训练网络拓扑

项目 3
RIP 的配置

3.1 用户需求

某学校网络拓扑如图 3-1 所示,要求完成路由信息协议(Routing Information Protocol,RIP)的配置,实现计算机 PC1、PC2 和 PC3 互通。

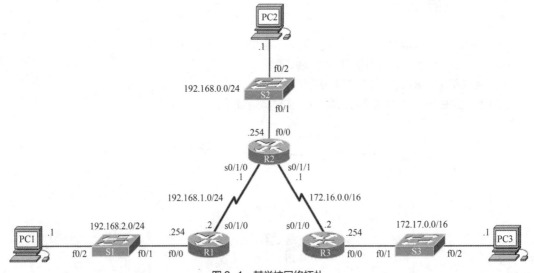

图 3-1 某学校网络拓扑

3.2 知识梳理

3.2.1 动态路由协议及其分类

动态路由协议是在路由器之间交换路由信息的协议,各路由器可以通过动态路由协议动态共享有关远程网络的信息,并自动将信息添加到各自的路由表中。动态路由协议可以确定到达各个网络的最佳路径,然后将最佳路径添加到路由表中。若网络拓扑结构发生了变化,那么路由器可通过交换路由信息自动获知新增加的网络,还可在当前网络连接出现故障时找出备用路径。

动态路由协议及其分类

根据路由协议的特性，动态路由协议可以分为内部网关协议（Interior Gateway Protocol，IGP）和外部网关协议（Exterior Gateway Protocol，EGP）。内部网关协议是用于自治系统内部的路由协议，外部网关协议是用于自治系统之间的路由协议。自治系统（Autonomous System，AS）也称为路由域，是指位于共同管理区域内的一组路由器。内部网关协议可以划分为距离矢量路由协议和链路状态路由协议。常用的距离矢量路由协议有 RIP、增强型内部网关路由协议（Enhanced Interior Gateway Routing Protocol，EIGRP），常用的链路状态路由协议有开放最短路径优先（Open Shortest Path First，OSPF）协议和中间系统到中间系统（Intermediate System to Intermediate System，IS-IS）协议。常用的外部网关协议有边界网关协议（Border Gateway Protocol，BGP）。

（1）距离矢量路由协议

距离矢量路由协议将路由器作为通往最终目的地的路径上的路标。启用了距离矢量路由协议的路由器并不了解整个网络的拓扑结构，路由器了解的是远程网络到该网络的距离（度量）及可以到达该网络的路径或接口。

（2）链路状态路由协议

启用了链路状态路由协议的路由器使用链路状态信息来创建网络拓扑，并在这些拓扑中选择到达所有目的网络的最佳路径。

根据路由器在传递路由信息时是否发送子网掩码，动态路由协议可以分为有类路由协议和无类路由协议。

（1）有类路由协议

有类路由协议在路由信息更新过程中不携带子网掩码信息。子网掩码是根据 A 类、B 类和 C 类 IP 地址的默认子网掩码来确定的。有类路由协议不支持可变长子网掩码（Variable Length Subnet Mask，VLSM）和不连续子网。

（2）无类路由协议

无类路由协议在路由信息更新过程中携带网络地址和子网掩码信息。无类路由协议支持 VLSM 和不连续子网。

3.2.2 RIP 的特点

1. RIPv1 和 RIPv2 的共同特点

（1）均为距离矢量路由协议。
（2）均使用跳数作为路径选择的唯一标准。
（3）通告的路由最大跳数为 15 跳，超过 15 跳会将该路由视为不可达。
（4）每 30s 广播一次消息。
（5）管理距离值是 120。

2. RIPv1 的局限

RIPv1 是一种有类路由协议，不支持 VLSM 和不连续子网，使用广播发送路由更新信息，广播地址是 255.255.255.255。

3. 与 RIPv1 相比，RIPv2 的增强特性

（1）是无类路由协议，支持 VLSM 和不连续子网。
（2）使用多播发送更新信息，多播地址为 224.0.0.9。
（3）支持对所有接口上的网络地址信息的手动路由汇总。
（4）支持身份验证机制以保证邻居之间路由表安全更新。

RIP 的特点和被动接口

> **边学边思**
>
> RIP在从RIPv1发展到RIPv2的过程中经历了一系列的优化，这些优化对于提升传输质量起到了积极的作用。作为国家发展的一分子，我们应该以质量强国的理念为指导，将其延伸到个人的成长与发展中，持续不断地提升自己的能力并积累知识，在各方面都追求精益求精，提升领域内的竞争力和影响力。

3.2.3 被动接口

被动接口可以阻止路由更新信息通过某个路由器接口传输，但仍然允许向其他路由器通告该路由更新信息。若启用 RIP 的某个接口连接的是网络末端，这时如果向这个接口通告路由更新信息，就会把带宽浪费在传输不必要的路由更新信息上，路由更新信息可能会被数据包嗅探软件中途截取，给网络带来安全方面的威胁。要解决上述问题，通常可以把连接网络末端的接口配置成被动接口。

3.2.4 RIP 的配置

1. RIPv1 的配置

（1）进入路由器 RIP 配置模式

```
Router(config)#router rip
```

该命令并不直接启动 RIP 进程，而是在配置了 RIP 的情况下提供对路由器配置模式的访问。

（2）通告指定网络

```
Router(config-router)#network network-address
```

network-address：每个直连网络的有类网络地址。在特定网络所属的所有接口上启用 RIP，接口开始发送和接收路由更新信息，在每 30s 一次的路由更新中向其他路由器通告该指定网络。

2. RIPv2 的配置

RIPv2 的配置过程中，进入路由器 RIP 配置模式和通告网络的方式与 RIPv1 的配置相同，只是 RIPv2 的配置中需要增加两条命令。

（1）把 RIP 的版本修改为第 2 版

```
Router(config-router)#version 2
```

（2）禁用自动汇总

```
Router(config-router)#no auto-summary
```

3. 被动接口的配置

```
Router(config-router)#passive-interface type number
```

type number：接口的类型和编号。

4. 显示路由器当前配置的路由协议

```
Router#show ip protocols
```

5. 显示 RIP 数据库信息

```
Router#show ip rip database
```

3.2.5 路由环路的避免

路由环路是指数据包在一系列路由器之间不断传输却始终无法到达其预期目的网络的一种现象。当两台或多台路由器的路由信息中存在指向不可达目的网络的有效路径时，就可能发生路由环路。

如图 3-2 所示，路由器 R1 直连了 10.0.0.0 和 10.1.0.0 网络，路由器 R2 直连了 10.1.0.0 和 10.2.0.0 网络，路由器 R3 直连了 10.2.0.0 和 10.3.0.0 网络，3 台路由器运行 RIP 收敛后，每台路由器都有对应 4 个网络的 4 条路由。如果路由器 R3 直连的 10.3.0.0 网络的 f0/0 接口发生了故障（不可用），因为直连路由出现的条件是接口必须处于 up/up 状态，所以路由器 R3 的路由表中 10.3.0.0 网络的路由会被删除。因为 RIP 会定期更新，每隔 30s 发送一次路由更新信息，所以 10.3.0.0 网络无效的路由更新信息不会被立刻发送给路由器 R2，必须等更新时间到了，才能发送路由更新信息。如果此时路由器 R2 的更新时间比路由器 R3 的先到了，那么路由器 R2 会把从路由器 R3 获取的 10.3.0.0 网络度量值为 1 的路由发送给路由器 R3，路由器 R3 收到后会把 10.3.0.0 网络的路由写入路由表并把度量值加 1（度量值为 2），到了下一个更新时间，路由器 R3 会把路由更新信息再发给路由器 R2，路由器 R2 中 10.3.0.0 网络的度量值会被更新为 3。路由器 R2 会认为要把数据包发给 10.3.0.0 网络，需要把数据包发给路由器 R3，路由器 R3 会认为要把数据包发给 10.3.0.0 网络，需要把数据包发给路由器 R2，当路由器 R2 收到一个目的地址是 10.3.0.0 网络的数据包时，在路由器 R2 和 R3 之间会发生路由环路。

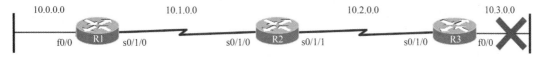

路由表		
目的网络	接口	度量值
10.0.0.0	f0/0	0
10.1.0.0	s0/1/0	0
10.2.0.0	s0/1/0	1
10.3.0.0	s0/1/0	2

路由表		
目的网络	接口	度量值
10.1.0.0	s0/1/0	0
10.2.0.0	s0/1/1	0
10.0.0.0	s0/1/0	1
10.3.0.0	s0/1/1	1

路由表		
目的网络	接口	度量值
10.2.0.0	s0/1/0	0
10.3.0.0	f0/0	0
10.1.0.0	s0/1/0	1
10.0.0.0	s0/1/0	2

图 3-2 路由环路

避免路由环路的方法有以下几种。

1. 触发更新

造成路由环路的原因是 RIP 收敛缓慢，更新的路由信息没有被及时发送给其他路由器。触发更新是指当路由表发生变化时不必等到更新时间的到来，而是立刻发送路由更新信息给其他路由器。图 3-2 中，当路由器 R3 直连的 10.3.0.0 网络的 f0/0 接口发生故障时，网络拓扑结构发生了变化，触发更新会使路由器 R3 直连的 10.3.0.0 网络无效的路由更新信息被立刻发送给其他路由器。

2. 设置度量值

距离矢量路由协议通过指定一个度量值来限定无穷大。一旦路由器计数达到该"无穷大"值，该路由就会被标记为不可达，RIP 将无穷大定义为 16 跳。图 3-2 中，当路由器 R3 直连的 10.3.0.0 网络的 f0/0 接口发生故障时，路由器 R3 中目的网络为 10.3.0.0 的度量值为 2，路由器 R2 中目的网络为 10.3.0.0 的度量值为 3，随着 RIP 的定期更新，路由器 R2 把路由更新信息再发给路由器 R1，这样路由器 R1 中 10.3.0.0 网络的度量值会被更新为 4，经过多次更新后，度量值会不断增加，当度量值达到 16 时，该路由会被标记为不可达。

3. 水平分割

水平分割规则规定，路由器不能将相同网络的路由信息与接收更新信息相同的接口进行通告。对图 3-2 中的网络应用水平分割后，路由器 R2 中 10.3.0.0 网络的路由信息是从 s0/1/1 接口由路由器 R3 通告过来的，这样 10.3.0.0 网络的路由信息不允许再从路由器 R2 的 s0/1/1 接口通告出去。

4. 路由毒化

路由毒化可在发往其他路由器的路由更新信息中将路由标记为不可达。将路由标记为不可达的方法是将度量值设置为"无穷大"。RIP 毒化路由的"无穷大"值为 16。图 3-2 中，路由器 R3 直连的 10.3.0.0 网络的 f0/0 接口发生了故障，10.3.0.0 网络的路由会被标记为不可达，之后 10.3.0.0 网络的路由更新信息会被发送给其他路由器，这样比等待跳数达到"无穷大"的收敛速度更快。

5. 毒性逆转

毒性逆转是指若路由器从某一接口获取毒化路由，则可以忽略水平分割从同一接口通告毒化路由。图 3-2 中，路由器 R3 直连的 10.3.0.0 网络的 f0/0 接口发生了故障，10.3.0.0 网络的路由会被标记为不可达，路由器 R2 从 s0/1/1 接口收到了路由器 R3 通告的毒化路由，路由器 R2 可以把毒化路由从 s0/1/1 接口发送给路由器 R3。

6. 抑制计时器

抑制计时器可以用来防止定期更新信息错误地恢复某条可能已经发生故障的路由。抑制计时器通过以下方式工作：路由器从邻居处接收更新信息，该更新信息表明以前可以访问的网络现在已不可访问。路由器将该网络标记为 possibly down 并启动抑制计时器。如果在抑制期间从任何相邻路由器接收到含有更小度量值的有关该网络的更新信息，则恢复该网络并删除抑制计时器。如果在抑制期间从相邻路由器收到的更新信息包含的度量值与之前的相同或比之前的更大，则该更新信息将被忽略。此方法减少了路由的浮动，提高了网络的稳定性。

7. 存活时间

IP 报头中，存活时间（Time To Live，TTL）限制了数据包在被丢弃之前能够在网络中传输的跳数。设置 TTL 字段的目的是防止数据包无休止地在网络中传输。数据包的源设备会为 TTL 字段设置一个值，在到达目的地的过程中，每经过一台路由器，TTL 值就会减 1。如果 TTL 值被减为零，那么路由器将丢弃该数据包。

3.2.6 路由表

1. 路由表条目的组成

（1）直连条目

如图 3-3 所示，直连条目由 3 部分组成：路由来源、目的网络和送出接口。路由表中，第一列是路由来源，路由来源用于确定路由的获取方式。直连接口有两个路由来源代码。"C"用于确定直连路由，当某个接口配置了 IP 地址并被激活时，将会自动创建直连路由。"L"表示这是本地路由，当某个接口配置了 IP 地址并被激活时，会自动创建本地路由。不同 IOS 版本的路由器，直连条目是不同的。如果路由器用的是 IOS 15 之前的版本，那么路由表中的直连条目只有路由来源为"C"的直连路由条目，没有路由来源为"L"的本地路由条目。如果路由器用的是 IOS 15 之后的版本，那么直连条目中路由来源为"C"的直连路由条目和路由来源为"L"的本地路由条目都有。"192.168.0.0/24 is directly connected"是目的网络，目的网络包含远程网络的地址和该网络的连接方式。"FastEthernet0/0"是送出接口，送出接口是将数据包转发至目的网络时要使用的出接口。

```
C    192.168.0.0/24 is directly connected, FastEthernet0/0
L    192.168.0.254/32 is directly connected, FastEthernet0/0
```

图 3-3 直连条目

（2）远程网络路由条目

如图 3-4 所示，远程网络路由条目由 7 部分组成：路由来源、目的网络、管理距离值、度量值、下一跳、路由时间戳和送出接口。路由表中，第一列的"R"是路由来源，路由来源用于确定路由的获取方式，"R"表示使用 RIP 从另一台路由器动态获取的路由。"192.168.0.0/24"是目的网络，目的网络用于确定远程网络的地址。"[120/1]"中的"120"是管理距离值，管理距离值定义路由的优先级，用于确定路由来源的可靠性。"[120/1]"中的"1"是度量值，度量值用于确定到达远程网络的分配值，较小的值表示首选路由。"172.16.0.1"是下一跳，下一跳用于确定下一跳路由器的 IPv4 地址以转发数据包至该路由器。"00:00:19"是路由时间戳，路由时间戳用于确定最后一次侦听路由的时间。"Serial0/1/0"是送出接口，送出接口用于确定将数据包转发至最终目的地的出接口。

```
R    192.168.0.0/24 [120/1] via 172.16.0.1, 00:00:19, Serial0/1/0
```

图 3-4 远程网络路由条目

2. 路由表结构

路由器的路由表采用分层结构，在查找路由并转发数据包时，这样的结构可加快查找速度。路由表中常用的路由术语主要有 1 级路由、最终路由、1 级父路由和 2 级子路由等。

（1）1 级路由

1 级路由是指子网掩码等于或小于有类掩码的路由，如图 3-5 所示，目的网络为 192.168.0.0/24 和 192.168.1.0/24 的路由都属于 1 级路由，因为它们的目的网络的子网掩码等于有类掩码。

1 级路由可被用作以下几种路由。

① 默认路由：指地址为 0.0.0.0/0 的静态路由。

② 超网路由：指子网掩码小于有类掩码的路由。

③ 网络路由：指子网掩码等于有类掩码的路由。网络路由也可以是 1 级父路由。

（2）最终路由

最终路由是包含下一跳 IPv4 地址或送出接口的路由，如图 3-5 所示，目的网络为 192.168.0.0/24 的路由包含下一跳，目的网络为 192.168.1.0/24 的路由包含送出接口，这两条路由都为最终路由。

```
S    192.168.0.0/24 [1/0] via 192.168.1.2
C    192.168.1.0/24 is directly connected, Serial0/1/0
```

图 3-5 1 级路由和最终路由

（3）1 级父路由和 2 级子路由

1 级父路由是指不包含任何网络的下一跳 IP 地址或送出接口的网络路由。1 级父路由实际上表示存在 2 级子路由的一个标题，2 级子路由是指有类网络地址的子网路由，2 级子路由也称为子路由，2 级子路由的来源可以是直连网络、静态路由或动态路由协议，如图 3-6 所示。

```
     172.16.0.0/23 is subnetted, 1 subnets
S       172.16.0.0 [1/0] via 192.168.1.1
```

图 3-6 1 级父路由和 2 级子路由

由图 3-6 可知，1 级父路由为

```
172.16.0.0/23 is subnetted, 1 subnets
```

2 级子路由为

```
S 172.16.0.0 [1/0] via 192.168.1.1
```

如果所有的 2 级子路由的网络前缀都相同,那么网络前缀会显示在 1 级父路由中;如果 2 级子路由的网络前缀不同,那么 1 级父路由中将不会出现网络前缀,每条 2 级子路由都有网络前缀。

3.3 方案设计

图 3-1 所示的网络拓扑中,3 台路由器 R1、R2 和 R3 互联了 5 个网络,路由器 R1 有 3 个非直连网络,路由器 R2 有两个非直连网络,路由器 R3 有 3 个非直连网络。因为路由器 R1、R2 和 R3 没有去往非直连网络的路由,所以网络无法互通。要实现网络的互通,可以为路由器 R1、R2 和 R3 配置 RIP。

3.4 项目实施

3.4.1 RIPv1 的配置

本小节网络拓扑如图 3-1 所示,路由器 R1、R2 和 R3 的接口,以及计算机 PC1、PC2 和 PC3 的 IP 地址已经配置完成,要求完成 RIPv1 的配置,实现计算机 PC1、PC2 和 PC3 互通。

步骤 1:在路由器 R1 的全局配置模式下配置 RIPv1,输入以下代码。

```
R1(config)#router rip
R1(config-router)#network 192.168.1.0
R1(config-router)#network 192.168.2.0
```

步骤 2:在路由器 R2 的全局配置模式下配置 RIPv1,输入以下代码。

```
R2(config)#router rip
R2(config-router)#network 192.168.0.0
R2(config-router)#network 192.168.1.0
R2(config-router)#network 172.16.0.0
```

步骤 3:在路由器 R3 的全局配置模式下配置 RIPv1,输入以下代码。

```
R3(config)#router rip
R3(config-router)#network 172.16.0.0
R3(config-router)#network 172.17.0.0
```

步骤 4:在路由器 R1 的特权执行模式下输入 show ip route,查看路由表,如图 3-7 所示。

```
R1#show ip route
Codes: C - connected, S - static, R - RIP, M - mobile, B - BGP
       D - EIGRP, EX - EIGRP external, O - OSPF, IA - OSPF inter area
       N1 - OSPF NSSA external type 1, N2 - OSPF NSSA external type 2
       E1 - OSPF external type 1, E2 - OSPF external type 2
       i - IS-IS, su - IS-IS summary, L1 - IS-IS level-1, L2 - IS-IS level-2
       ia - IS-IS inter area, * - candidate default, U - per-user static route
       o - ODR, P - periodic downloaded static route

Gateway of last resort is not set

R    172.17.0.0/16 [120/2] via 192.168.1.1, 00:00:01, Serial0/1/0
R    172.16.0.0/16 [120/1] via 192.168.1.1, 00:00:01, Serial0/1/0
R    192.168.0.0/24 [120/1] via 192.168.1.1, 00:00:01, Serial0/1/0
C    192.168.1.0/24 is directly connected, Serial0/1/0
C    192.168.2.0/24 is directly connected, FastEthernet0/0
```

图 3-7 路由器 R1 的路由表

步骤 5：在路由器 R2 的特权执行模式下输入 show ip route，查看路由表，如图 3-8 所示。

```
R2#show ip route
Codes: C - connected, S - static, R - RIP, M - mobile, B - BGP
       D - EIGRP, EX - EIGRP external, O - OSPF, IA - OSPF inter area
       N1 - OSPF NSSA external type 1, N2 - OSPF NSSA external type 2
       E1 - OSPF external type 1, E2 - OSPF external type 2
       i - IS-IS, su - IS-IS summary, L1 - IS-IS level-1, L2 - IS-IS level-2
       ia - IS-IS inter area, * - candidate default, U - per-user static route
       o - ODR, P - periodic downloaded static route

Gateway of last resort is not set

R    172.17.0.0/16 [120/1] via 172.16.0.2, 00:00:19, Serial0/1/1
C    172.16.0.0/16 is directly connected, Serial0/1/1
C    192.168.0.0/24 is directly connected, FastEthernet0/0
C    192.168.1.0/24 is directly connected, Serial0/1/0
R    192.168.2.0/24 [120/1] via 192.168.1.2, 00:00:26, Serial0/1/0
```

图 3-8　路由器 R2 的路由表

步骤 6：在路由器 R3 的特权执行模式下输入 show ip route，查看路由表，如图 3-9 所示。

```
R3#show ip route
Codes: C - connected, S - static, R - RIP, M - mobile, B - BGP
       D - EIGRP, EX - EIGRP external, O - OSPF, IA - OSPF inter area
       N1 - OSPF NSSA external type 1, N2 - OSPF NSSA external type 2
       E1 - OSPF external type 1, E2 - OSPF external type 2
       i - IS-IS, su - IS-IS summary, L1 - IS-IS level-1, L2 - IS-IS level-2
       ia - IS-IS inter area, * - candidate default, U - per-user static route
       o - ODR, P - periodic downloaded static route

Gateway of last resort is not set

C    172.17.0.0/16 is directly connected, FastEthernet0/0
C    172.16.0.0/16 is directly connected, Serial0/1/0
R    192.168.0.0/24 [120/1] via 172.16.0.1, 00:00:19, Serial0/1/0
R    192.168.1.0/24 [120/1] via 172.16.0.1, 00:00:19, Serial0/1/0
R    192.168.2.0/24 [120/2] via 172.16.0.1, 00:00:19, Serial0/1/0
```

图 3-9　路由器 R3 的路由表

步骤 7：在计算机 PC1 的命令行界面输入 ping 192.168.0.1，检验联通性，如图 3-10 所示。

```
C:\>ping 192.168.0.1

正在 Ping 192.168.0.1 具有 32 字节的数据：
来自 192.168.0.1 的回复：字节=32 时间=9ms TTL=126
来自 192.168.0.1 的回复：字节=32 时间=9ms TTL=126
来自 192.168.0.1 的回复：字节=32 时间=9ms TTL=126
来自 192.168.0.1 的回复：字节=32 时间=9ms TTL=126

192.168.0.1 的 Ping 统计信息：
    数据包：已发送 = 4，已接收 = 4，丢失 = 0 (0% 丢失)，
往返行程的估计时间(以毫秒为单位)：
    最短 = 9ms，最长 = 9ms，平均 = 9ms
```

图 3-10　从计算机 PC1 ping 计算机 PC2

步骤 8：在计算机 PC1 的命令行界面输入 ping 172.17.0.1，检验联通性，如图 3-11 所示。

```
C:\>ping 172.17.0.1

正在 Ping 172.17.0.1 具有 32 字节的数据：
来自 172.17.0.1 的回复：字节=32 时间=19ms TTL=125
来自 172.17.0.1 的回复：字节=32 时间=18ms TTL=125
来自 172.17.0.1 的回复：字节=32 时间=18ms TTL=125
来自 172.17.0.1 的回复：字节=32 时间=18ms TTL=125

172.17.0.1 的 Ping 统计信息：
    数据包：已发送 = 4，已接收 = 4，丢失 = 0 (0% 丢失)，
往返行程的估计时间(以毫秒为单位)：
    最短 = 18ms，最长 = 19ms，平均 = 18ms
```

图 3-11　从计算机 PC1 ping 计算机 PC3

步骤 9：在计算机 PC2 的命令行界面输入 ping 172.17.0.1，检验联通性，如图 3-12 所示。

```
C:\>ping 172.17.0.1
正在 Ping 172.17.0.1 具有 32 字节的数据:
来自 172.17.0.1 的回复: 字节=32 时间=9ms TTL=126
来自 172.17.0.1 的回复: 字节=32 时间=9ms TTL=126
来自 172.17.0.1 的回复: 字节=32 时间=9ms TTL=126
来自 172.17.0.1 的回复: 字节=32 时间=9ms TTL=126
172.17.0.1 的 Ping 统计信息:
    数据包: 已发送 = 4, 已接收 = 4, 丢失 = 0 (0% 丢失),
往返行程的估计时间(以毫秒为单位):
    最短 = 9ms, 最长 = 9ms, 平均 = 9ms
```

图 3-12　从计算机 PC2 ping 计算机 PC3

通过查看路由表和检验联通性可以发现，RIPv1 动态路由在路由表中是路由来源为 R、管理距离值为 120 的路由，路由表中的路由完整，计算机 PC1、PC2 和 PC3 实现了互通。

3.4.2　RIPv1 不连续子网的配置

本小节网络拓扑如图 3-13 所示，路由器 R1、R2 和 R3 的接口，以及计算机 PC1、PC2 和 PC3 的 IP 地址已经配置完成，要求完成 RIPv1 不连续子网的配置，检验计算机 PC1、PC2 和 PC3 的联通性。

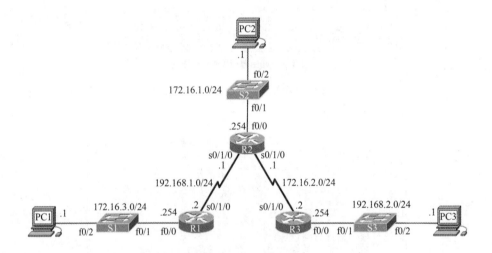

图 3-13　网络拓扑

步骤 1：在路由器 R1 的全局配置模式下配置 RIPv1，输入以下代码。

```
R1(config)#router rip
R1(config-router)#network 192.168.1.0
R1(config-router)#network 172.16.0.0
```

步骤 2：在路由器 R2 的全局配置模式下配置 RIPv1，输入以下代码。

```
R2(config)#router rip
R2(config-router)#network 172.16.0.0
R2(config-router)#network 192.168.1.0
```

步骤 3：在路由器 R3 的全局配置模式下配置 RIPv1，输入以下代码。

```
R3(config)#router rip
R3(config-router)#network 172.16.0.0
R3(config-router)#network 192.168.2.0
```

步骤 4：在路由器 R1 的特权执行模式下输入 show ip route，查看路由表，如图 3-14 所示。

```
R1#show ip route
Codes: C - connected, S - static, R - RIP, M - mobile, B - BGP
       D - EIGRP, EX - EIGRP external, O - OSPF, IA - OSPF inter area
       N1 - OSPF NSSA external type 1, N2 - OSPF NSSA external type 2
       E1 - OSPF external type 1, E2 - OSPF external type 2
       i - IS-IS, su - IS-IS summary, L1 - IS-IS level-1, L2 - IS-IS level-2
       ia - IS-IS inter area, * - candidate default, U - per-user static route
       o - ODR, P - periodic downloaded static route

Gateway of last resort is not set

     172.16.0.0/24 is subnetted, 1 subnets
C       172.16.3.0 is directly connected, FastEthernet0/0
C    192.168.1.0/24 is directly connected, Serial0/1/0
R    192.168.2.0/24 [120/2] via 192.168.1.1, 00:00:24, Serial0/1/0
```

图 3-14　路由器 R1 的路由表

步骤 5：在路由器 R2 的特权执行模式下输入 show ip route，查看路由表，如图 3-15 所示。

```
R2#show ip route
Codes: C - connected, S - static, R - RIP, M - mobile, B - BGP
       D - EIGRP, EX - EIGRP external, O - OSPF, IA - OSPF inter area
       N1 - OSPF NSSA external type 1, N2 - OSPF NSSA external type 2
       E1 - OSPF external type 1, E2 - OSPF external type 2
       i - IS-IS, su - IS-IS summary, L1 - IS-IS level-1, L2 - IS-IS level-2
       ia - IS-IS inter area, * - candidate default, U - per-user static route
       o - ODR, P - periodic downloaded static route

Gateway of last resort is not set

     172.16.0.0/24 is subnetted, 2 subnets
C       172.16.1.0 is directly connected, FastEthernet0/0
C       172.16.2.0 is directly connected, Serial0/1/1
C    192.168.1.0/24 is directly connected, Serial0/1/0
R    192.168.2.0/24 [120/1] via 172.16.2.2, 00:00:10, Serial0/1/1
```

图 3-15　路由器 R2 的路由表

步骤 6：在路由器 R3 的特权执行模式下输入 show ip route，查看路由表，如图 3-16 所示。

```
R3#show ip route
Codes: C - connected, S - static, R - RIP, M - mobile, B - BGP
       D - EIGRP, EX - EIGRP external, O - OSPF, IA - OSPF inter area
       N1 - OSPF NSSA external type 1, N2 - OSPF NSSA external type 2
       E1 - OSPF external type 1, E2 - OSPF external type 2
       i - IS-IS, su - IS-IS summary, L1 - IS-IS level-1, L2 - IS-IS level-2
       ia - IS-IS inter area, * - candidate default, U - per-user static route
       o - ODR, P - periodic downloaded static route

Gateway of last resort is not set

     172.16.0.0/24 is subnetted, 2 subnets
R       172.16.1.0 [120/1] via 172.16.2.1, 00:00:15, Serial0/1/0
C       172.16.2.0 is directly connected, Serial0/1/0
R    192.168.1.0/24 [120/1] via 172.16.2.1, 00:00:15, Serial0/1/0
C    192.168.2.0/24 is directly connected, FastEthernet0/0
```

图 3-16　路由器 R3 的路由表

步骤 7：在计算机 PC1 的命令行界面输入 ping 172.16.1.1，检验联通性，如图 3-17 所示。

```
C:\>ping 172.16.1.1
正在 Ping 172.16.1.1 具有 32 字节的数据:
来自 172.16.3.254 的回复: 无法访问目标主机。
来自 172.16.3.254 的回复: 无法访问目标主机。
来自 172.16.3.254 的回复: 无法访问目标主机。
来自 172.16.3.254 的回复: 无法访问目标主机。

172.16.1.1 的 Ping 统计信息:
    数据包: 已发送 = 4，已接收 = 4，丢失 = 0 (0% 丢失)，
```

图 3-17　从计算机 PC1 ping 计算机 PC2

步骤 8：在计算机 PC1 的命令行界面输入 ping 192.168.2.1，检验联通性，如图 3-18 所示。

```
C:\>ping 192.168.2.1
正在 Ping 192.168.2.1 具有 32 字节的数据:
请求超时。
请求超时。
请求超时。
请求超时。

192.168.2.1 的 Ping 统计信息:
    数据包: 已发送 = 4，已接收 = 0，丢失 = 4 (100% 丢失)，
```

图 3-18　从计算机 PC1 ping 计算机 PC3

步骤 9：在计算机 PC2 的命令行界面输入 ping 192.168.2.1，检验联通性，如图 3-19 所示。

```
C:\>ping 192.168.2.1
正在 Ping 192.168.2.1 具有 32 字节的数据:
来自 192.168.2.1 的回复: 字节=32 时间=9ms TTL=126
来自 192.168.2.1 的回复: 字节=32 时间=9ms TTL=126
来自 192.168.2.1 的回复: 字节=32 时间=9ms TTL=126
来自 192.168.2.1 的回复: 字节=32 时间=9ms TTL=126

192.168.2.1 的 Ping 统计信息:
    数据包: 已发送 = 4，已接收 = 4，丢失 = 0 (0% 丢失)，
    往返行程的估计时间(以毫秒为单位):
        最短 = 9ms，最长 = 9ms，平均 = 9ms
```

图 3-19　从计算机 PC2 ping 计算机 PC3

通过查看路由表和检验联通性可以发现，路由器 R1、R2 和 R3 的路由表中的路由不完整，没有实现计算机 PC1、PC2 和 PC3 互通。因为 RIPv1 不支持不连续子网，所以针对图 3-13 中 IP 地址属于不连续子网，RIPv1 无法实现网络互通。

3.4.3　RIPv2 的配置

本小节网络拓扑如图 3-13 所示，路由器 R1、R2 和 R3 的接口，以及计算机 PC1、PC2 和 PC3 的 IP 地址已经配置完成，要求完成 RIPv2 的配置，实现计算机 PC1、PC2 和 PC3 互通。

步骤 1：在路由器 R1 的全局配置模式下配置 RIPv2，输入以下代码。

```
R1(config)#router rip
R1(config-router)#version 2
```

```
R1(config-router)#network 192.168.1.0
R1(config-router)#network 172.16.0.0
R1(config-router)#no auto-summary
```

步骤 2：在路由器 R2 的全局配置模式下配置 RIPv2，输入以下代码。

```
R2(config)#router rip
R2(config-router)#version 2
R2(config-router)#network 172.16.0.0
R2(config-router)#network 192.168.1.0
R2(config-router)#no auto-summary
```

步骤 3：在路由器 R3 的全局配置模式下配置 RIPv2，输入以下代码。

```
R3(config)#router rip
R3(config-router)#version 2
R3(config-router)#network 172.16.0.0
R3(config-router)#network 192.168.2.0
R3(config-router)#no auto-summary
```

步骤 4：在路由器 R1 的特权执行模式下输入 show ip route，查看路由表，如图 3-20 所示。

```
R1#show ip route
Codes: C - connected, S - static, R - RIP, M - mobile, B - BGP
       D - EIGRP, EX - EIGRP external, O - OSPF, IA - OSPF inter area
       N1 - OSPF NSSA external type 1, N2 - OSPF NSSA external type 2
       E1 - OSPF external type 1, E2 - OSPF external type 2
       i - IS-IS, su - IS-IS summary, L1 - IS-IS level-1, L2 - IS-IS level-2
       ia - IS-IS inter area, * - candidate default, U - per-user static route
       o - ODR, P - periodic downloaded static route

Gateway of last resort is not set

     172.16.0.0/16 is variably subnetted, 4 subnets, 2 masks
R       172.16.0.0/16 is possibly down,
          routing via 192.168.1.1, Serial0/1/0
R       172.16.1.0/24 [120/1] via 192.168.1.1, 00:00:09, Serial0/1/0
R       172.16.2.0/24 [120/1] via 192.168.1.1, 00:00:09, Serial0/1/0
C       172.16.3.0/24 is directly connected, FastEthernet0/0
C    192.168.1.0/24 is directly connected, Serial0/1/0
R    192.168.2.0/24 [120/2] via 192.168.1.1, 00:00:09, Serial0/1/0
```

图 3-20　路由器 R1 的路由表

步骤 5：在路由器 R2 的特权执行模式下输入 show ip route，查看路由表，如图 3-21 所示。

```
R2#show ip route
Codes: C - connected, S - static, R - RIP, M - mobile, B - BGP
       D - EIGRP, EX - EIGRP external, O - OSPF, IA - OSPF inter area
       N1 - OSPF NSSA external type 1, N2 - OSPF NSSA external type 2
       E1 - OSPF external type 1, E2 - OSPF external type 2
       i - IS-IS, su - IS-IS summary, L1 - IS-IS level-1, L2 - IS-IS level-2
       ia - IS-IS inter area, * - candidate default, U - per-user static route
       o - ODR, P - periodic downloaded static route

Gateway of last resort is not set

     172.16.0.0/24 is subnetted, 3 subnets
C       172.16.1.0 is directly connected, FastEthernet0/0
C       172.16.2.0 is directly connected, Serial0/1/1
R       172.16.3.0 [120/1] via 192.168.1.2, 00:00:11, Serial0/1/0
C    192.168.1.0/24 is directly connected, Serial0/1/0
R    192.168.2.0/24 [120/1] via 172.16.2.2, 00:00:02, Serial0/1/1
```

图 3-21　路由器 R2 的路由表

步骤 6：在路由器 R3 的特权执行模式下输入 show ip route，查看路由表，如图 3-22 所示。

```
R3#show ip route
Codes: C - connected, S - static, R - RIP, M - mobile, B - BGP
       D - EIGRP, EX - EIGRP external, O - OSPF, IA - OSPF inter area
       N1 - OSPF NSSA external type 1, N2 - OSPF NSSA external type 2
       E1 - OSPF external type 1, E2 - OSPF external type 2
       i - IS-IS, su - IS-IS summary, L1 - IS-IS level-1, L2 - IS-IS level-2
       ia - IS-IS inter area, * - candidate default, U - per-user static route
       o - ODR, P - periodic downloaded static route

Gateway of last resort is not set

     172.16.0.0/24 is subnetted, 3 subnets
R       172.16.1.0 [120/1] via 172.16.2.1, 00:00:21, Serial0/1/0
C       172.16.2.0 is directly connected, Serial0/1/0
R       172.16.3.0 [120/2] via 172.16.2.1, 00:00:21, Serial0/1/0
R    192.168.1.0/24 [120/1] via 172.16.2.1, 00:00:21, Serial0/1/0
C    192.168.2.0/24 is directly connected, FastEthernet0/0
```

图 3-22 路由器 R3 的路由表

步骤 7：在计算机 PC1 的命令行界面输入 ping 172.16.1.1，检验联通性，如图 3-23 所示。

```
C:\>ping 172.16.1.1

正在 Ping 172.16.1.1 具有 32 字节的数据:
来自 172.16.1.1 的回复: 字节=32 时间=10ms TTL=126
来自 172.16.1.1 的回复: 字节=32 时间=9ms TTL=126
来自 172.16.1.1 的回复: 字节=32 时间=9ms TTL=126
来自 172.16.1.1 的回复: 字节=32 时间=9ms TTL=126

172.16.1.1 的 Ping 统计信息:
    数据包: 已发送 = 4, 已接收 = 4, 丢失 = 0 (0% 丢失),
往返行程的估计时间(以毫秒为单位):
    最短 = 9ms, 最长 = 10ms, 平均 = 9ms
```

图 3-23 从计算机 PC1 ping 计算机 PC2

步骤 8：在计算机 PC1 的命令行界面输入 ping 192.168.2.1，检验联通性，如图 3-24 所示。

```
C:\>ping 192.168.2.1

正在 Ping 192.168.2.1 具有 32 字节的数据:
来自 192.168.2.1 的回复: 字节=32 时间=18ms TTL=125
来自 192.168.2.1 的回复: 字节=32 时间=19ms TTL=125
来自 192.168.2.1 的回复: 字节=32 时间=18ms TTL=125
来自 192.168.2.1 的回复: 字节=32 时间=18ms TTL=125

192.168.2.1 的 Ping 统计信息:
    数据包: 已发送 = 4, 已接收 = 4, 丢失 = 0 (0% 丢失),
往返行程的估计时间(以毫秒为单位):
    最短 = 18ms, 最长 = 19ms, 平均 = 18ms
```

图 3-24 从计算机 PC1 ping 计算机 PC3

步骤 9：在计算机 PC2 的命令行界面输入 ping 192.168.2.1，检验联通性，如图 3-25 所示。

```
C:\>ping 192.168.2.1

正在 Ping 192.168.2.1 具有 32 字节的数据:
来自 192.168.2.1 的回复: 字节=32 时间=9ms TTL=126
来自 192.168.2.1 的回复: 字节=32 时间=9ms TTL=126
来自 192.168.2.1 的回复: 字节=32 时间=9ms TTL=126
来自 192.168.2.1 的回复: 字节=32 时间=9ms TTL=126

192.168.2.1 的 Ping 统计信息:
    数据包: 已发送 = 4, 已接收 = 4, 丢失 = 0 (0% 丢失),
往返行程的估计时间(以毫秒为单位):
    最短 = 9ms, 最长 = 9ms, 平均 = 9ms
```

图 3-25 从计算机 PC2 ping 计算机 PC3

通过查看路由表和检验联通性可以发现，路由器 R1、R2 和 R3 的路由表中路由完整，计算机 PC1、PC2 和 PC3 实现了互通。因为 RIPv2 是一种无类路由协议，支持不连续子网，所以针对图 3-13 中 IP 地址属于不连续子网，可以通过配置 RIPv2 实现网络互通。

3.4.4 被动接口的配置

本小节网络拓扑如图 3-13 所示，路由器的接口和计算机的 IP 地址已经配置完成，为路由器 R1、R2 和 R3 配置了 RIPv2，已经实现了全网互通，要求完成被动接口的配置，减少不必要的路由通告流量。

步骤 1：在路由器 R1 的全局配置模式下配置被动接口，输入以下代码。

```
R1(config)#router rip
R1(config-router)#passive-interface f0/0
```

步骤 2：在路由器 R2 的全局配置模式下配置被动接口，输入以下代码。

```
R2(config)#router rip
R2(config-router)#passive-interface f0/0
```

步骤 3：在路由器 R3 的全局配置模式下配置被动接口，输入以下代码。

```
R3(config)#router rip
R3(config-router)#passive-interface f0/0
```

3.5 项目小结

本项目完成了 RIPv1 和 RIPv2 的配置。RIPv1 是一种有类路由协议，不支持 VLSM 和不连续子网，所以对于 IP 地址属于不连续子网，如果为路由器配置 RIPv1，则无法实现全网互通。对于这样的网络，为路由器配置 RIPv2 便可以实现全网互通。

3.6 拓展训练

本项目的拓展训练网络拓扑如图 3-26 所示，要求完成如下配置：
（1）完成路由器接口和计算机 IP 地址的配置；
（2）完成 RIPv2 的配置，实现计算机 PC1、PC2 和 PC3 互通；
（3）要求业务网段中不出现协议报文。

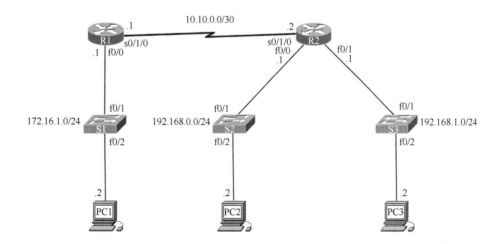

图 3-26 项目 3 拓展训练网络拓扑

项目 4
EIGRP的配置

4.1 用户需求

某学校网络拓扑如图 4-1 所示,要求完成 EIGRP 的配置,实现计算机 PC1、PC2 和 PC3 互通。

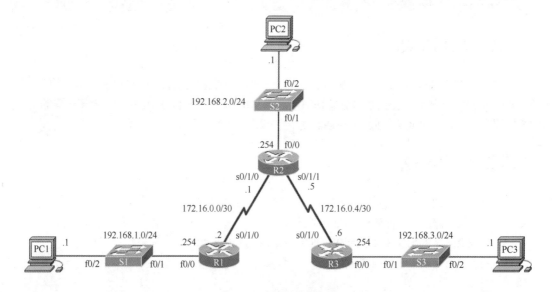

图 4-1 某学校网络拓扑

4.2 知识梳理

4.2.1 EIGRP 的特点

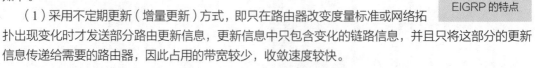

EIGRP 的特点

EIGRP 是思科的专有协议,它是一种距离矢量路由协议。EIGRP 主要特点如下。

(1)采用不定期更新(增量更新)方式,即只在路由器改变度量标准或网络拓扑出现变化时才发送部分路由更新信息,更新信息中只包含变化的链路信息,并且只将这部分的更新信息传递给需要的路由器,因此占用的带宽较少,收敛速度较快。

(2)路由更新信息中包含子网掩码,支持 VLSM 和不连续子网。

（3）采用多播（224.0.0.10）或单播进行路由更新。

（4）内部 EIGRP 路由的默认管理距离值为 90，从外部（如静态路由）导入的 EIGRP 路由的默认管理距离值为 170。

（5）对每一种网络协议，EIGRP 都维持独立的邻居表、拓扑表和路由表。拓扑表包含最佳路由和所有无环备用路径。

（6）支持等价和非等价的负载均衡。

（7）使用扩散更新算法（Diffusing Update Algorithm，DUAL）来实现快速收敛，并确保没有路由环路。

（8）存储整个网络拓扑结构的信息，以便快速适应网络变化。

4.2.2 与 EIGRP 相关的术语

与 EIGRP 相关的术语

与 EIGRP 相关的术语如下。

1. 邻居表

使用 Hello 数据包来发现邻居，路由器发现新邻居并建立邻接关系后，将在邻居表中添加一个条目，其中包含该邻居的地址及可能到达邻居的接口，用于确保直连邻居之间能够进行双向通信。EIGRP 为支持的每种协议都维护一个邻居表。

2. 拓扑表

路由器动态发现新邻居后，将向它发送更新信息，其中包含自己知道的路由信息，同时此路由器也将收到来自其新邻居的更新信息，这些更新信息将用于填充拓扑表。每台路由器都将邻居的路由表信息存储在自己的拓扑表中，EIGRP 为支持的每种协议都维护一个拓扑表。

3. 可行距离

可行距离（Feasible Distance，FD）是指计算出的通向目的网络的最小度量值，是路由表条目中所列的度量值。可行距离是当前路由器到下一跳路由器的开销加上下一跳路由器到目的地的开销。

4. 通告距离

通告距离（Advertised Distance，AD）是邻居通向相同目的网络的可行距离，是路由器向邻居通告的、有关自身通向该网络的开销，是下一跳路由器到目的地的开销。

5. 后继路由器

后继路由器（Successor Router）是指用于转发数据包的一个邻居，它是通向目的网络的开销最小的路由器，是到达远端网络的最佳路由器，也是路由表中路由器的下一跳。通过它到达目的网络的路径是最优的。

6. 可行后继路由器

可行后继路由器（Feasible Successor Router）是指一个邻居，它有一条通向后继路由器所联通的同一个目的网络的无环备用路径，并且满足可行性条件。通过它到达目的地的度量值比后继路由器的大，但它的通告距离小于通过后继路由器到达目的网络的可行距离，因而该路由器被保存在拓扑表中，用作备份。

7. 路由表

路由表包含前往每个目的地的最佳路径信息，用于转发数据包，前往后继路由器的路由信息被存储到路由表中。路由器为配置的每种网络协议都维护一个路由表，默认情况下，每种协议最多可以将 4 条前往同一个目的地且度量值相同的路由加入路由表中。

4.2.3 EIGRP 的数据包类型

EIGRP 的数据包类型主要有 5 种，分别如下。

EIGRP 的数据包类型

1. Hello 数据包

Hello 数据包用来发现和维护 EIGRP 邻居关系，以多播方式发送，目的地址为 224.0.0.10。

在大多数网络中，一般每 5s 发送一次 Hello 数据包。在多点非广播多路访问（Non-Broadcast Multiple Access，NBMA）网络中，以及在 X.25、帧中继和带有 T1（1.544 Mbit/s）或更慢访问链路的异步传输模式（Asynchronous Transfer Mode，ATM）接口上，一般每 60s 单播一次 Hello 数据包。EIGRP 路由器假定：只要还能收到邻居发来的 Hello 数据包，该邻居及其路由表就仍然保持活动状态。

保持时间用于告诉路由器在通告邻居无法到达前应等待该设备发送下一个 Hello 数据包的最长时间。默认情况下，保持时间是该设备发送 Hello 数据包间隔时长的 3 倍。当保持时间截止后，EIGRP 将通告该路由器发生故障，并从其邻居表中删除相应邻居的条目，同时清除与该邻居相关的路由信息。

2. Update 数据包

Update 数据包用于传播路由信息。EIGRP 仅在必要时才发送 Update 数据包。Update 数据包仅包含需要的路由信息，且仅被发送给需要该信息的路由器。EIGRP 的 Update 数据包使用可靠传输，邻居收到后必须回复确认消息。当多台路由器需要 Update 数据包时，通过多播发送 Update 数据包；当只有一台路由器需要 Update 数据包时，则通过单播发送 Update 数据包。

3. Query 数据包

当路由信息丢失且没有备份路径时，使用 Query 数据包向邻居发送查询信息，邻居必须回复确认信息。

4. Reply 数据包

Reply 数据包是对邻居发送的 Query 数据包的回复，也需要邻居回复确认消息。

5. Ack 数据包

Ack 数据包是 EIGRP 在使用可靠传输时发送的，包括对收到的数据包的确认消息，告诉邻居已经收到数据包了，收到 Ack 数据包后不需要再回复。

4.2.4 EIGRP 的度量值

EIGRP 使用的是复合度量值，包括带宽、延迟、可靠性和负载。默认情况下仅使用带宽和延迟。复合度量值的计算公式：

复合度量值 $=\left[k_1 \times 带宽 + (k_2 \times 带宽)/(256-负载) + k_3 \times 延迟\right] \times \left[k_5/(可靠性+k_4)\right]$

复合度量值的计算公式是一个条件公式，当 $k_5=0$ 时，公式中最后一部分 $\left[k_5/(可靠性+k_4)\right]$ 的值认定为 1。默认情况下，$k_1=1$，$k_2=0$，$k_3=1$，$k_4=0$，$k_5=0$，复合度量值的计算公式可以简化为：

复合度量值 $=\left[k_1 \times 带宽 + k_3 \times 延迟\right]$

4.2.5 配置命令

1. 启用 EIGRP

```
Router(config)# router eigrp autonomous-system
```

autonomous-system：自治系统编号，它实际上起进程 ID 的作用。EIGRP 路由域内的所有路由器都

必须使用同一个进程 ID。

2. 通告网络

```
Router(config-router)#network network-address [wildcard-mask]
```

network-address：接口的有类网络地址。

wildcard-mask：反掩码。有时，网络管理员并不想为所有接口启用 EIGRP，此时要配置 EIGRP 仅通告特定子网，可以将 *wildcard-mask* 选项与 network 命令结合使用。

3. 禁用自动汇总

```
Router(config-router)#no auto-summary
```

4. 手动汇总

```
Router(config-if)#ip summary-address eigrp as-number network-address subnet-mask
```

5. 被动接口的配置

```
Router(config-router)#passive-interface type number
```

type number：接口的类型和编号。

6. 显示路由器当前配置的路由协议

```
Router#show ip protocols
```

7. 查看邻居表并检验是否已经与邻居建立邻接关系

```
Router#show ip eigrp neighbors
```

8. 查看拓扑表

```
Router#show ip eigrp topology [all-links]
```

添加 all-links 选项可以查看所有可能的链路。

4.3 方案设计

图 4-1 所示的网络拓扑中，3 台路由器 R1、R2 和 R3 互联了 5 个网络，路由器 R1 有 3 个非直连网络，路由器 R2 有两个非直连网络，路由器 R3 有 3 个非直连网络，因为路由器 R1、R2 和 R3 没有去往非直连网络的路由，所以网络无法互通。要实现网络的互通，可以为路由器 R1、R2 和 R3 配置 EIGRP。

4.4 项目实施

4.4.1 一般情况下 EIGRP 的配置

本小节网络拓扑如图 4-1 所示，路由器 R1、R2 和 R3 的接口，以及计算机 PC1、PC2 和 PC3 的 IP 地址已经配置完成，要求完成一般情况下 EIGRP 的配置，实现计算机 PC1、PC2 和 PC3 互通。

步骤 1：在路由器 R1 的全局配置模式下配置 EIGRP，输入以下代码。

```
R1(config)#router eigrp 1
R1(config-router)#network 192.168.1.0
R1(config-router)#network 172.16.0.0
R1(config-router)#no auto-summary
```

步骤 2：在路由器 R2 的全局配置模式下配置 EIGRP，输入以下代码。

```
R2(config)#router eigrp 1
R2(config-router)#network 192.168.2.0
R2(config-router)#network 172.16.0.0
R2(config-router)#no auto-summary
```

步骤 3：在路由器 R3 的全局配置模式下配置 EIGRP，输入以下代码。

```
R3(config)#router eigrp 1
R3(config-router)#network 172.16.0.0
R3(config-router)#network 192.168.3.0
R3(config-router)#no auto-summary
```

步骤 4：在路由器 R1 的特权执行模式下输入 show ip route，查看路由表，如图 4-2 所示。

```
R1#show ip route
Codes: C - connected, S - static, R - RIP, M - mobile, B - BGP
       D - EIGRP, EX - EIGRP external, O - OSPF, IA - OSPF inter area
       N1 - OSPF NSSA external type 1, N2 - OSPF NSSA external type 2
       E1 - OSPF external type 1, E2 - OSPF external type 2
       i - IS-IS, su - IS-IS summary, L1 - IS-IS level-1, L2 - IS-IS level-2
       ia - IS-IS inter area, * - candidate default, U - per-user static route
       o - ODR, P - periodic downloaded static route

Gateway of last resort is not set

     172.16.0.0/30 is subnetted, 2 subnets
D       172.16.0.4 [90/21024000] via 172.16.0.1, 00:06:48, Serial0/1/0
C       172.16.0.0 is directly connected, Serial0/1/0
C    192.168.1.0/24 is directly connected, FastEthernet0/0
D    192.168.2.0/24 [90/20514560] via 172.16.0.1, 00:06:48, Serial0/1/0
D    192.168.3.0/24 [90/21026560] via 172.16.0.1, 00:05:50, Serial0/1/0
```

图 4-2　路由器 R1 的路由表

步骤 5：在路由器 R2 的特权执行模式下输入 show ip route，查看路由表，如图 4-3 所示。

```
R2#show ip route
Codes: C - connected, S - static, R - RIP, M - mobile, B - BGP
       D - EIGRP, EX - EIGRP external, O - OSPF, IA - OSPF inter area
       N1 - OSPF NSSA external type 1, N2 - OSPF NSSA external type 2
       E1 - OSPF external type 1, E2 - OSPF external type 2
       i - IS-IS, su - IS-IS summary, L1 - IS-IS level-1, L2 - IS-IS level-2
       ia - IS-IS inter area, * - candidate default, U - per-user static route
       o - ODR, P - periodic downloaded static route

Gateway of last resort is not set

     172.16.0.0/30 is subnetted, 2 subnets
C       172.16.0.4 is directly connected, Serial0/1/1
C       172.16.0.0 is directly connected, Serial0/1/0
D    192.168.1.0/24 [90/20514560] via 172.16.0.2, 00:05:38, Serial0/1/0
C    192.168.2.0/24 is directly connected, FastEthernet0/0
D    192.168.3.0/24 [90/20514560] via 172.16.0.6, 00:04:40, Serial0/1/1
```

图 4-3　路由器 R2 的路由表

步骤 6：在路由器 R3 的特权执行模式下输入 show ip route，查看路由表，如图 4-4 所示。

```
R3#show ip route
Codes: C - connected, S - static, R - RIP, M - mobile, B - BGP
       D - EIGRP, EX - EIGRP external, O - OSPF, IA - OSPF inter area
       N1 - OSPF NSSA external type 1, N2 - OSPF NSSA external type 2
       E1 - OSPF external type 1, E2 - OSPF external type 2
       i - IS-IS, su - IS-IS summary, L1 - IS-IS level-1, L2 - IS-IS level-2
       ia - IS-IS inter area, * - candidate default, U - per-user static route
       o - ODR, P - periodic downloaded static route

Gateway of last resort is not set

     172.16.0.0/30 is subnetted, 2 subnets
C       172.16.0.4 is directly connected, Serial0/1/0
D       172.16.0.0 [90/21024000] via 172.16.0.5, 00:03:29, Serial0/1/0
D    192.168.1.0/24 [90/21026560] via 172.16.0.5, 00:03:29, Serial0/1/0
D    192.168.2.0/24 [90/20514560] via 172.16.0.5, 00:03:29, Serial0/1/0
C    192.168.3.0/24 is directly connected, FastEthernet0/0
```

图 4-4　路由器 R3 的路由表

步骤7：在计算机 PC1 的命令行界面输入 ping 192.168.2.1，检验联通性，如图 4-5 所示。

```
C:\>ping 192.168.2.1

正在 Ping 192.168.2.1 具有 32 字节的数据:
来自 192.168.2.1 的回复: 字节=32 时间=9ms TTL=126
来自 192.168.2.1 的回复: 字节=32 时间=9ms TTL=126
来自 192.168.2.1 的回复: 字节=32 时间=9ms TTL=126
来自 192.168.2.1 的回复: 字节=32 时间=9ms TTL=126

192.168.2.1 的 Ping 统计信息:
    数据包: 已发送 = 4，已接收 = 4，丢失 = 0 (0% 丢失)，
往返行程的估计时间(以毫秒为单位):
    最短 = 9ms，最长 = 9ms，平均 = 9ms
```

图 4-5 从计算机 PC1 ping 计算机 PC2

步骤8：在计算机 PC1 的命令行界面输入 ping 192.168.3.1，检验联通性，如图 4-6 所示。

```
C:\>ping 192.168.3.1

正在 Ping 192.168.3.1 具有 32 字节的数据:
来自 192.168.3.1 的回复: 字节=32 时间=18ms TTL=125
来自 192.168.3.1 的回复: 字节=32 时间=18ms TTL=125
来自 192.168.3.1 的回复: 字节=32 时间=19ms TTL=125
来自 192.168.3.1 的回复: 字节=32 时间=19ms TTL=125

192.168.3.1 的 Ping 统计信息:
    数据包: 已发送 = 4，已接收 = 4，丢失 = 0 (0% 丢失)，
往返行程的估计时间(以毫秒为单位):
    最短 = 18ms，最长 = 19ms，平均 = 18ms
```

图 4-6 从计算机 PC1 ping 计算机 PC3

步骤9：在计算机 PC2 的命令行界面输入 ping 192.168.3.1，检验联通性，如图 4-7 所示。

```
C:\>ping 192.168.3.1

正在 Ping 192.168.3.1 具有 32 字节的数据:
来自 192.168.3.1 的回复: 字节=32 时间=9ms TTL=126
来自 192.168.3.1 的回复: 字节=32 时间=9ms TTL=126
来自 192.168.3.1 的回复: 字节=32 时间=9ms TTL=126
来自 192.168.3.1 的回复: 字节=32 时间=9ms TTL=126

192.168.3.1 的 Ping 统计信息:
    数据包: 已发送 = 4，已接收 = 4，丢失 = 0 (0% 丢失)，
往返行程的估计时间(以毫秒为单位):
    最短 = 9ms，最长 = 9ms，平均 = 9ms
```

图 4-7 从计算机 PC2 ping 计算机 PC3

通过查看路由表和检验联通性可以发现，EIGRP 在路由表中是路由来源为"D"、管理距离值为 90 的路由，路由器 R1、R2 和 R3 的路由表中的路由完整，计算机 PC1、PC2 和 PC3 实现了互通。

4.4.2 禁用自动汇总时 EIGRP 的配置

本小节网络拓扑如图 4-8 所示，路由器 R1、R2 和 R3 的接口，以及计算机 PC1、PC2 和 PC3 的 IP 地址已经配置完成，要求完成禁用自动汇总时 EIGRP 的配置，实现计算机 PC1、PC2 和 PC3 互通，并观察路由器 R1、R2 和 R3 的路由表。

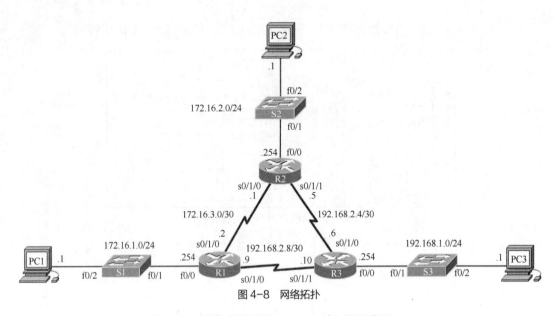

图 4-8 网络拓扑

步骤 1：在路由器 R1 的全局配置模式下配置 EIGRP，输入以下代码。

```
R1(config)#router eigrp 1
R1(config-router)#network 172.16.1.0 0.0.0.255
R1(config-router)#network 172.16.3.0 0.0.0.3
R1(config-router)#network 192.168.2.8 0.0.0.3
```

步骤 2：在路由器 R2 的全局配置模式下配置 EIGRP，输入以下代码。

```
R2(config)#router eigrp 1
R2(config-router)#network 172.16.2.0 0.0.0.255
R2(config-router)#network 172.16.3.0 0.0.0.3
R2(config-router)#network 192.168.2.4 0.0.0.3
```

步骤 3：在路由器 R3 的全局配置模式下配置 EIGRP，输入以下代码。

```
R3(config)#router eigrp 1
R3(config-router)#network 192.168.1.0
R3(config-router)#network 192.168.2.4 0.0.0.3
R3(config-router)#network 192.168.2.8 0.0.0.3
```

步骤 4：在路由器 R1 的特权执行模式下输入 show ip route，查看路由表，如图 4-9 所示。

```
R1#show ip route
Codes: C - connected, S - static, R - RIP, M - mobile, B - BGP
       D - EIGRP, EX - EIGRP external, O - OSPF, IA - OSPF inter area
       N1 - OSPF NSSA external type 1, N2 - OSPF NSSA external type 2
       E1 - OSPF external type 1, E2 - OSPF external type 2
       i - IS-IS, su - IS-IS summary, L1 - IS-IS level-1, L2 - IS-IS level-2
       ia - IS-IS inter area, * - candidate default, U - per-user static route
       o - ODR, P - periodic downloaded static route

Gateway of last resort is not set

     172.16.0.0/16 is variably subnetted, 4 subnets, 3 masks
D       172.16.0.0/16 is a summary, 00:04:37, Null0
C       172.16.1.0/24 is directly connected, FastEthernet0/0
D       172.16.2.0/24 [90/20514560] via 172.16.3.1, 00:04:08, Serial0/1/0
C       172.16.3.0/30 is directly connected, Serial0/1/0
D    192.168.1.0/24 [90/20514560] via 192.168.2.10, 00:07:51, Serial0/1/1
     192.168.2.0/24 is variably subnetted, 3 subnets, 2 masks
C       192.168.2.8/30 is directly connected, Serial0/1/1
D       192.168.2.0/24 is a summary, 00:07:53, Null0
D       192.168.2.4/30 [90/21024000] via 192.168.2.10, 00:07:51, Serial0/1/1
```

图 4-9 路由器 R1 的路由表

步骤 5：在路由器 R2 的特权执行模式下输入 show ip route，查看路由表，如图 4-10 所示。

```
R2#show ip route
Codes: C - connected, S - static, R - RIP, M - mobile, B - BGP
       D - EIGRP, EX - EIGRP external, O - OSPF, IA - OSPF inter area
       N1 - OSPF NSSA external type 1, N2 - OSPF NSSA external type 2
       E1 - OSPF external type 1, E2 - OSPF external type 2
       i - IS-IS, su - IS-IS summary, L1 - IS-IS level-1, L2 - IS-IS level-2
       ia - IS-IS inter area, * - candidate default, U - per-user static route
       o - ODR, P - periodic downloaded static route

Gateway of last resort is not set

     172.16.0.0/16 is variably subnetted, 4 subnets, 3 masks
D       172.16.0.0/16 is a summary, 00:02:53, Null0
D       172.16.1.0/24 [90/20514560] via 172.16.3.2, 00:03:21, Serial0/1/0
C       172.16.2.0/24 is directly connected, FastEthernet0/0
C       172.16.3.0/30 is directly connected, Serial0/1/0
D    192.168.1.0/24 [90/20514560] via 192.168.2.6, 00:06:35, Serial0/1/1
     192.168.2.0/24 is variably subnetted, 3 subnets, 2 masks
D       192.168.2.8/30 [90/21024000] via 192.168.2.6, 00:07:28, Serial0/1/1
D       192.168.2.0/24 is a summary, 00:06:37, Null0
C       192.168.2.4/30 is directly connected, Serial0/1/1
```

图 4-10　路由器 R2 的路由表

步骤 6：在路由器 R3 的特权执行模式下输入 show ip route，查看路由表，如图 4-11 所示。

```
R3#show ip route
Codes: C - connected, S - static, R - RIP, M - mobile, B - BGP
       D - EIGRP, EX - EIGRP external, O - OSPF, IA - OSPF inter area
       N1 - OSPF NSSA external type 1, N2 - OSPF NSSA external type 2
       E1 - OSPF external type 1, E2 - OSPF external type 2
       i - IS-IS, su - IS-IS summary, L1 - IS-IS level-1, L2 - IS-IS level-2
       ia - IS-IS inter area, * - candidate default, U - per-user static route
       o - ODR, P - periodic downloaded static route

Gateway of last resort is not set

D    172.16.0.0/16 [90/20514560] via 192.168.2.9, 00:05:34, Serial0/1/1
                   [90/20514560] via 192.168.2.5, 00:05:34, Serial0/1/0
C    192.168.1.0/24 is directly connected, FastEthernet0/0
     192.168.2.0/24 is variably subnetted, 3 subnets, 2 masks
C       192.168.2.8/30 is directly connected, Serial0/1/1
D       192.168.2.0/24 is a summary, 00:11:09, Null0
C       192.168.2.4/30 is directly connected, Serial0/1/0
```

图 4-11　路由器 R3 的路由表

步骤 7：在计算机 PC1 的命令行界面输入 ping 172.16.2.1，检验联通性，如图 4-12 所示。

```
C:\>ping 172.16.2.1

正在 Ping 172.16.2.1 具有 32 字节的数据：
来自 172.16.2.1 的回复：字节=32 时间=9ms TTL=126
来自 172.16.2.1 的回复：字节=32 时间=9ms TTL=126
来自 172.16.2.1 的回复：字节=32 时间=9ms TTL=126
来自 172.16.2.1 的回复：字节=32 时间=9ms TTL=126

172.16.2.1 的 Ping 统计信息：
    数据包：已发送 = 4，已接收 = 4，丢失 = 0 (0% 丢失)，
往返行程的估计时间(以毫秒为单位)：
    最短 = 9ms，最长 = 9ms，平均 = 9ms
```

图 4-12　从计算机 PC1 ping 计算机 PC2

步骤 8：在计算机 PC1 的命令行界面输入 ping 192.168.1.1，检验联通性，如图 4-13 所示。

```
C:\>ping 192.168.1.1
正在 Ping 192.168.1.1 具有 32 字节的数据:
来自 192.168.1.1 的回复: 字节=32 时间=14ms TTL=125
来自 192.168.1.1 的回复: 字节=32 时间=14ms TTL=125
来自 192.168.1.1 的回复: 字节=32 时间=14ms TTL=125
来自 192.168.1.1 的回复: 字节=32 时间=14ms TTL=125

192.168.1.1 的 Ping 统计信息:
    数据包: 已发送 = 4,已接收 = 4,丢失 = 0 (0% 丢失),
往返行程的估计时间(以毫秒为单位):
    最短 = 14ms,最长 = 14ms,平均 = 14ms
```

图 4-13　从计算机 PC1 ping 计算机 PC3

步骤 9:在计算机 PC2 的命令行界面输入 ping 192.168.1.1,检验联通性,如图 4-14 所示。

```
C:\>ping 192.168.1.1
正在 Ping 192.168.1.1 具有 32 字节的数据:
来自 192.168.1.1 的回复: 字节=32 时间=9ms TTL=126
来自 192.168.1.1 的回复: 字节=32 时间=9ms TTL=126
来自 192.168.1.1 的回复: 字节=32 时间=9ms TTL=126
来自 192.168.1.1 的回复: 字节=32 时间=9ms TTL=126

192.168.1.1 的 Ping 统计信息:
    数据包: 已发送 = 4,已接收 = 4,丢失 = 0 (0% 丢失),
往返行程的估计时间(以毫秒为单位):
    最短 = 9ms,最长 = 9ms,平均 = 9ms
```

图 4-14　从计算机 PC2 ping 计算机 PC3

步骤 10:在路由器 R1 的全局配置模式下禁用自动汇总,输入以下代码。

```
R1(config)#router eigrp 1
R1(config-router)#no auto-summary
```

步骤 11:在路由器 R2 的全局配置模式下禁用自动汇总,输入以下代码。

```
R2(config)#router eigrp 1
R2(config-router)#no auto-summary
```

步骤 12:在路由器 R3 的全局配置模式下禁用自动汇总,输入以下代码。

```
R3(config)#router eigrp 1
R3(config-router)#no auto-summary
```

步骤 13:在路由器 R1 的特权执行模式下输入 show ip route,查看路由表,如图 4-15 所示。

```
R1#show ip route
Codes: C - connected, S - static, R - RIP, M - mobile, B - BGP
       D - EIGRP, EX - EIGRP external, O - OSPF, IA - OSPF inter area
       N1 - OSPF NSSA external type 1, N2 - OSPF NSSA external type 2
       E1 - OSPF external type 1, E2 - OSPF external type 2
       i - IS-IS, su - IS-IS summary, L1 - IS-IS level-1, L2 - IS-IS level-2
       ia - IS-IS inter area, * - candidate default, U - per-user static route
       o - ODR, P - periodic downloaded static route

Gateway of last resort is not set

     172.16.0.0/16 is variably subnetted, 3 subnets, 2 masks
C       172.16.1.0/24 is directly connected, FastEthernet0/0
D       172.16.2.0/24 [90/20514560] via 172.16.3.1, 00:00:26, Serial0/1/0
C       172.16.3.0/30 is directly connected, Serial0/1/0
D    192.168.1.0/24 [90/20514560] via 192.168.2.10, 00:12:08, Serial0/1/1
     192.168.2.0/30 is subnetted, 2 subnets
C       192.168.2.8 is directly connected, Serial0/1/1
D       192.168.2.4 [90/21024000] via 192.168.2.10, 00:00:26, Serial0/1/1
                    [90/21024000] via 172.16.3.1, 00:00:26, Serial0/1/0
```

图 4-15　路由器 R1 的路由表

步骤 14：在路由器 R2 的特权执行模式下输入 show ip route，查看路由表，如图 4-16 所示。

```
R2#show ip route
Codes: C - connected, S - static, R - RIP, M - mobile, B - BGP
       D - EIGRP, EX - EIGRP external, O - OSPF, IA - OSPF inter area
       N1 - OSPF NSSA external type 1, N2 - OSPF NSSA external type 2
       E1 - OSPF external type 1, E2 - OSPF external type 2
       i - IS-IS, su - IS-IS summary, L1 - IS-IS level-1, L2 - IS-IS level-2
       ia - IS-IS inter area, * - candidate default, U - per-user static route
       o - ODR, P - periodic downloaded static route

Gateway of last resort is not set

     172.16.0.0/16 is variably subnetted, 3 subnets, 2 masks
D       172.16.1.0/24 [90/20514560] via 172.16.3.2, 00:01:15, Serial0/1/0
C       172.16.2.0/24 is directly connected, FastEthernet0/0
C       172.16.3.0/30 is directly connected, Serial0/1/0
D    192.168.1.0/24 [90/20514560] via 192.168.2.6, 00:13:17, Serial0/1/1
     192.168.2.0/30 is subnetted, 2 subnets
D       192.168.2.8 [90/21024000] via 192.168.2.6, 00:01:15, Serial0/1/1
                    [90/21024000] via 172.16.3.2, 00:01:15, Serial0/1/0
C       192.168.2.4 is directly connected, Serial0/1/1
```

图 4-16　路由器 R2 的路由表

步骤 15：在路由器 R3 的特权执行模式下输入 show ip route，查看路由表，如图 4-17 所示。

```
R3#show ip route
Codes: C - connected, S - static, R - RIP, M - mobile, B - BGP
       D - EIGRP, EX - EIGRP external, O - OSPF, IA - OSPF inter area
       N1 - OSPF NSSA external type 1, N2 - OSPF NSSA external type 2
       E1 - OSPF external type 1, E2 - OSPF external type 2
       i - IS-IS, su - IS-IS summary, L1 - IS-IS level-1, L2 - IS-IS level-2
       ia - IS-IS inter area, * - candidate default, U - per-user static route
       o - ODR, P - periodic downloaded static route

Gateway of last resort is not set

     172.16.0.0/16 is variably subnetted, 3 subnets, 2 masks
D       172.16.1.0/24 [90/20514560] via 192.168.2.9, 00:02:46, Serial0/1/1
D       172.16.2.0/24 [90/20514560] via 192.168.2.5, 00:02:46, Serial0/1/0
D       172.16.3.0/30 [90/21024000] via 192.168.2.9, 00:02:46, Serial0/1/1
                      [90/21024000] via 192.168.2.5, 00:02:46, Serial0/1/0
C    192.168.1.0/24 is directly connected, FastEthernet0/0
     192.168.2.0/30 is subnetted, 2 subnets
C       192.168.2.8 is directly connected, Serial0/1/1
C       192.168.2.4 is directly connected, Serial0/1/0
```

图 4-17　路由器 R3 的路由表

通过查看路由表和检验联通性可以发现，在没有禁用自动汇总之前，路由器 R1、R2 和 R3 的路由表中有 Null0 汇总路由，禁用自动汇总后，Null0 汇总路由被删除，路由器 R1、R2 和 R3 的路由表中的路由完整，网络实现了互通。

4.5　项目小结

本项目完成了 EIGRP 的配置。EIGRP 是思科的专有协议，默认情况下，EIGRP 会自动对路由进行汇总，在路由器的路由表中添加 Null0 汇总路由。Null0 汇总路由实际上是不通向任何地方的路由，EIGRP 使用 Null0 汇总路由来丢弃与 1 级父路由匹配但与所有 2 级子路由都不匹配的数据包。禁用自动汇总会删除 Null0 汇总路由并允许 EIGRP 在 2 级子路由与目的数据包不匹配时寻找超网路由或默认路由。

4.6　拓展训练

本项目的拓展训练网络拓扑如图 4-18 所示，要求完成如下配置：

（1）完成路由器接口和计算机 IP 地址的配置；
（2）完成 EIGRP 的配置，实现计算机 PC1、PC2、PC3 和 PC4 互通；
（3）要求业务网段中不出现协议报文。

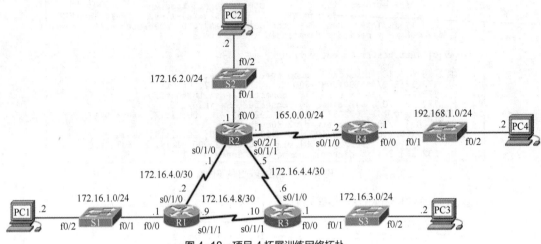

图 4-18　项目 4 拓展训练网络拓扑

项目 5
OSPF 的配置

5.1 用户需求

某学校网络拓扑如图 5-1 所示,要求完成 OSPF 的配置,实现计算机 PC1、PC2 和 PC3 互通。

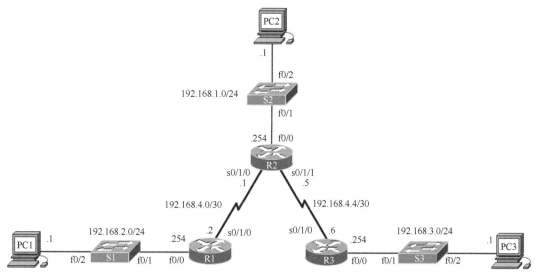

图 5-1 某学校网络拓扑

5.2 知识梳理

5.2.1 OSPF 的特点

OSPF 的特点主要如下。
(1) OSPF 是链路状态路由协议。
(2) OSPF 是真正的无路由环路的路由协议。
(3) OSPF 收敛速度快,能够在最短的时间内将路由变化传递到整个自治系统。
(4) 提出区域 (Area) 划分的概念,将自治系统划分为不同区域后,通过区域之间对路由信息的摘要,大大减少了需要传递的路由信息数量,也使得路由信

OSPF 的特点

息不会随网络规模的扩大而急剧膨胀。

（5）在广播网络中，使用多播地址发送路由更新信息和控制信息，减少对其他不运行 OSPF 的网络设备的干扰。

（6）OSPF 采用开销（Cost）作为度量值。

（7）OSPF 支持 MD5 身份验证。启用该功能后，OSPF 路由器只接收来自对等设备中具有相同预共享密钥的加密路由更新。

（8）OSPF 适应各种规模的网络，最多可支持数千台计算机。

（9）OSPF 默认的管理距离值是 110。

5.2.2 OSPF 的数据包类型

OSPF 的数据包类型主要有 5 种，分别如下。

1. Hello 数据包

Hello 数据包用于与其他 OSPF 路由器建立和维持邻接关系。Hello 数据包用于发现 OSPF 邻居，并建立邻接关系；通告两台路由器建立邻接关系所必须统一的参数；在以太网和帧中继网络等多路访问网络中选举指定路由器（Designated Router，DR）和备用指定路由器（Backup Designated Router，BDR）。

默认情况下，在多路访问网段和点对点网段中，每 10s 发送一次 Hello 数据包，在 NBMA 网段中则每 30s 发送一次 Hello 数据包。

路由器在通告邻居进入故障状态之前，等待该设备发送 Hello 数据包的时长是 Dead Time，单位为秒。思科所用的默认 Dead Time 为 Hello Time 的 4 倍。对于多路访问网段和点对点网段，此时长为 40s，对于 NBMA 网段则为 120s。

2. DBD 数据包

数据库描述（Database Description，DBD）数据包包含发送方路由器的链路状态数据库的简略列表，用于接收方路由器检查本地链路状态数据库。同一区域的所有链路状态路由器上的链路状态数据库必须保持一致。

3. LSR 数据包

接收方路由器可以通过发送链路状态请求（Link State Request，LSR）数据包来请求 DBD 数据包中任何条目的有关详细信息。

4. LSU 数据包

链路状态更新（Link State Update，LSU）数据包用于回复 LSR 数据包和通告新信息。

5. LSAck 数据包

路由器收到 LSU 数据包后，会发送一个链路状态确认（Link State Acknowledgment，LSAck）数据包来确认接收了 LSU 数据包。

5.2.3 OSPF Router ID

OSPF Router ID 用于唯一标识 OSPF 路由域内的每台路由器。一个 Router ID 其实就是一个 IP 地址。Router ID 通过以下步骤确定。

（1）使用通过 OSPF router-id 命令配置的 IP 地址。

（2）如果没有为路由器配置 Router ID，那么路由器会选择其所有环回口的最高 IP 地址（具有最大数值的 IP 地址）。

（3）如果没有为路由器配置 Router ID 和环回口，那么路由器会选择所有活动物理接口的最高 IP 地址。

5.2.4 OSPF 的网络类型

OSPF 的网络类型主要有 4 种，分别是 P2P、BMA、NBMA 和 P2MP，介绍如下。

1. P2P

点到点（Point-to-point，P2P）网络是将一对路由器连接起来的网络，当链路层协议是点到点协议（Point-to-Point Protocol，PPP）或高级数据链路控制（High-level Data Link Control，HDLC）协议时，OSPF 的默认网络类型是 P2P，这种类型的网络不进行 OSPF 的 DR 和 BDR 选举。

2. BMA

广播多路访问（Broadcast Multiple Access，BMA）网络支持广播，允许多台设备接入，任意两台设备都可以进行二层通信。当链路层协议是以太网（Ethernet）协议和光纤分布式数据接口（Fiber Distributed Data Interface，FDDI）协议时，OSPF 的默认网络类型是 BMA，这种类型的网络要进行 OSPF 的 DR 和 BDR 选举。

3. NBMA

NBMA 允许多台设备接入，但是不具备广播功能。当链路层协议是帧中继、ATM 和 X.25 协议时，OSPF 的默认网络类型是 NBMA。在 NBMA 网络中，也要进行 OSPF 的 DR 和 BDR 选举。

4. P2MP

在 OSPF 中，默认情况下，不会假定任何链路层协议对应的网络类型是点对多点（Point-to-Multipoint，P2MP）类型，这种类型的网络需要管理员手动配置，该网络不进行 OSPF 的 DR 和 BDR 选举。

5.2.5 DR 和 BDR

OSPF 多路访问网络会创建多边邻接关系，其中的每对路由器间都存在一项邻接关系，会造成链路状态通告（Link State Advertisement，LSA）数据包的大量泛洪，泛洪过程中的流量很大，将影响网络的正常通信。解决方法是，多路访问网络中的路由器选举出一个 DR 和一个 BDR，其他路由器变为非指定路由器（DR Other），DR 负责收集和分发 LSA，BDR 防止 DR 发生故障，DR Other 仅与网络中的 DR 和 BDR 建立完全的邻接关系。这样 DR Other 只需使用多播地址 224.0.0.6 将 LSA 发送给 DR 和 BDR 即可。

1. DR 和 BDR 的选举

（1）具有最高 OSPF 接口优先级的路由器当选为 DR。

（2）具有第二高 OSPF 接口优先级的路由器当选为 BDR。

（3）如果 OSPF 接口的优先级相同，则取 Router ID 最大者。

2. DR 和 BDR 选举的时间安排

当多路访问网络中第一台启用了 OSPF 接口的路由器开始工作时，DR 和 BDR 选举过程随即开始，一旦选出 DR，就保持 DR 的地位，直到出现下列条件之一为止。

（1）DR 发生故障。

（2）DR 上的 OSPF 进程发生故障。

（3）DR 上的多路访问接口发生故障。

5.2.6 虚链路

如果 OSPF 有多个区域，其中一个区域必须为骨干区域（area 0），其他所有区域都要与 area 0 直连，且 area 0 必须是连续的，为了避免路由环路，OSPF 所有的非骨干区域都要将路由通告给 area 0，以便将路由通告给其他区域。

虚链路可以将不连续的 area 0 连接起来，还可以把非骨干区域连接到 area 0。虚链路不能穿越多个区域，只能穿越标准的非骨干区域。如果需要使用虚链路穿越两个非骨干区域连接到 area 0，则需要两条虚链路。

5.2.7 度量值

OSPF 的度量值称为开销，开销与每个路由器接口的输出端关联，接口的开销通过计算 10 的 8 次幂（参考带宽）除以以 bit/s 为单位的带宽（默认带宽）得到。OSPF 规范规定了开销必须为整数，通常情况下，系统会将小于 1 的开销视为 1，以简化配置和管理。常用接口的开销如表 5-1 所示。

表 5-1 常用接口的开销

接口类型	参考带宽（bit/s）÷默认带宽（bit/s）	开销
10 千兆以太网 10 Gbit/s	100000000÷10000000000	1
千兆以太网 1 Gbit/s	100000000÷1000000000	1
快速以太网 100 Mbit/s	100000000÷100000000	1
以太网 10 Mbit/s	100000000÷10000000	10
串行 1.544 Mbit/s	100000000÷1544000	64
串行 128 kbit/s	100000000÷128000	781
串行 64 kbit/s	100000000÷64000	1562

5.2.8 配置命令

1. 启用 OSPF

```
Router(config)#router ospf process-id
```

process-id：一个介于 1 和 65535 之间的数字，由网络管理员选定。*process-id* 仅在本地有效，这意味着在路由器之间建立邻接关系时无须匹配该参数。

2. 通告网络

```
Router(config-router)#network network-address wildcard-mask area
 area-id
```

network-address：接口的网络地址，network 命令中网络地址的接口都将被启用，可发送和接收 OSPF 数据包，此网络（或子网）将被包括在 OSPF 路由更新信息中。

wildcard-mask：反掩码和网络地址，用于指定 network 命令启用的接口或接口范围。

area-id：OSPF 区域，OSPF 区域是共享链路状态信息的一组路由器。相同区域内的所有 OSPF 路由器的链路状态数据库中必须具有相同的链路状态信息，这通过路由器将各自的链路状态泛洪给该区域内的其他所有路由器来实现。如果所有路由器都处于同一个 OSPF 区域，则必须在所有路由器上使用相同的 *area-id* 来配置 network 命令，比较好的做法是在单区域 OSPF 中使用 area 0。

3. 配置 Router ID

```
Router(config-router)#router-id ip-address
```

4. 配置环回口

```
Router(config)#interface loopback number
Router(config-if)#ip address ip-address subnet-mask
```

5. 重新加载 OSPF 进程

```
Router#clear ip ospf process
```

6. DR 和 BDR 选举的控制

```
Router(config-if)#ip ospf priority value
```

value：优先级，取值范围为 0～255。

值为 0 表示该路由器不具备成为 DR 或 BDR 的资格。

1 是路由器默认的优先级的值。

7. 虚链路的配置

```
Router(config-router)#area area-id virtual-link router-id
```

area-id：指定虚链路经过的中转区域的区域 ID，可以是十进制数，也可以是类似于 IP 地址的点分十进制数，没有默认值。

router-id：指定虚链路另一端的路由器的 Router ID。

8. 被动接口的配置

```
Router(config-router)#passive-interface type number
```

type number：接口的类型和编号。

9. 显示路由器当前配置的路由协议

```
Router#show ip protocols
```

该命令可以查看路由器的 Router ID。

10. 显示邻居表

```
Router#show ip ospf neighbor [detail]
```

11. 查看接口是否被接入正确的区域

```
Router#show ip ospf interface
```

该命令还可以显示邻居关系和各种定时器信息，如 Hello Time 等。

12. 显示 OSPF 的 Router ID、定时器、最短通路优先算法的执行次数和 LSA 信息等

```
Router#show ip ospf
```

13. 查看 OSPF 数据库

```
Router#show ip ospf database
```

14. 查看 OSPF 的虚链路

```
Router#show ip ospf virtual-link
```

5.3 方案设计

图 5-1 所示的网络拓扑中，3 台路由器 R1、R2 和 R3 互联了 5 个网络，路由器 R1 有 3 个非直连网络，路由器 R2 有两个非直连网络，路由器 R3 有 3 个非直连网络，因为 R1、R2 和 R3 没有去往非直连网络的路由，所以网络无法互通。要实现网络的互通，可以为路由器 R1、R2 和 R3 配置 OSPF。

5.4 项目实施

5.4.1 单区域 OSPF 的配置

本小节网络拓扑如图 5-1 所示，路由器 R1、R2 和 R3 的接口，以及计算机 PC1、PC2 和 PC3 的 IP 地址已经配置完成，要求完成单区域 OSPF 的配置，实现计算机 PC1、PC2 和 PC3 互通。

步骤 1：在路由器 R1 的全局配置模式下配置 OSPF，输入以下代码。

```
R1(config)#router ospf 1
R1(config-router)#network 192.168.2.0 0.0.0.255 area 0
R1(config-router)#network 192.168.4.0 0.0.0.3 area 0
```

步骤 2：在路由器 R2 的全局配置模式下配置 OSPF，输入以下代码。

```
R2(config)#router ospf 1
R2(config-router)#network 192.168.1.0 0.0.0.255 area 0
R2(config-router)#network 192.168.4.0 0.0.0.3 area 0
R2(config-router)#network 192.168.4.4 0.0.0.3 area 0
```

步骤 3：在路由器 R3 的全局配置模式下配置 OSPF，输入以下代码。

```
R3(config)#router ospf 1
R3(config-router)#network 192.168.4.4 0.0.0.3 area 0
R3(config-router)#network 192.168.3.0 0.0.0.255 area 0
```

步骤 4：在路由器 R1 的特权执行模式下输入 show ip route，查看路由表，如图 5-2 所示。

```
R1#show ip route
Codes: C - connected, S - static, R - RIP, M - mobile, B - BGP
       D - EIGRP, EX - EIGRP external, O - OSPF, IA - OSPF inter area
       N1 - OSPF NSSA external type 1, N2 - OSPF NSSA external type 2
       E1 - OSPF external type 1, E2 - OSPF external type 2
       i - IS-IS, su - IS-IS summary, L1 - IS-IS level-1, L2 - IS-IS level-2
       ia - IS-IS inter area, * - candidate default, U - per-user static route
       o - ODR, P - periodic downloaded static route

Gateway of last resort is not set

     192.168.4.0/30 is subnetted, 2 subnets
O       192.168.4.4 [110/1562] via 192.168.4.1, 00:01:56, Serial0/1/0
C       192.168.4.0 is directly connected, Serial0/1/0
O    192.168.1.0/24 [110/782] via 192.168.4.1, 00:05:08, Serial0/1/0
C    192.168.2.0/24 is directly connected, FastEthernet0/0
O    192.168.3.0/24 [110/1563] via 192.168.4.1, 00:01:46, Serial0/1/0
```

图 5-2 路由器 R1 的路由表

步骤 5：在路由器 R2 的特权执行模式下输入 show ip route，查看路由表，如图 5-3 所示。

```
R2#show ip route
Codes: C - connected, S - static, R - RIP, M - mobile, B - BGP
       D - EIGRP, EX - EIGRP external, O - OSPF, IA - OSPF inter area
       N1 - OSPF NSSA external type 1, N2 - OSPF NSSA external type 2
       E1 - OSPF external type 1, E2 - OSPF external type 2
       i - IS-IS, su - IS-IS summary, L1 - IS-IS level-1, L2 - IS-IS level-2
       ia - IS-IS inter area, * - candidate default, U - per-user static route
       o - ODR, P - periodic downloaded static route

Gateway of last resort is not set

     192.168.4.0/30 is subnetted, 2 subnets
C       192.168.4.4 is directly connected, Serial0/1/1
C       192.168.4.0 is directly connected, Serial0/1/0
C    192.168.1.0/24 is directly connected, FastEthernet0/0
O    192.168.2.0/24 [110/782] via 192.168.4.2, 00:04:02, Serial0/1/0
O    192.168.3.0/24 [110/782] via 192.168.4.6, 00:00:50, Serial0/1/1
```

图 5-3　路由器 R2 的路由表

步骤 6：在路由器 R3 的特权执行模式下输入 show ip route，查看路由表，如图 5-4 所示。

```
R3#show ip route
Codes: C - connected, S - static, R - RIP, M - mobile, B - BGP
       D - EIGRP, EX - EIGRP external, O - OSPF, IA - OSPF inter area
       N1 - OSPF NSSA external type 1, N2 - OSPF NSSA external type 2
       E1 - OSPF external type 1, E2 - OSPF external type 2
       i - IS-IS, su - IS-IS summary, L1 - IS-IS level-1, L2 - IS-IS level-2
       ia - IS-IS inter area, * - candidate default, U - per-user static route
       o - ODR, P - periodic downloaded static route

Gateway of last resort is not set

     192.168.4.0/30 is subnetted, 2 subnets
C       192.168.4.4 is directly connected, Serial0/1/0
O       192.168.4.0 [110/1562] via 192.168.4.5, 00:02:56, Serial0/1/0
O    192.168.1.0/24 [110/782] via 192.168.4.5, 00:02:56, Serial0/1/0
O    192.168.2.0/24 [110/1563] via 192.168.4.5, 00:02:56, Serial0/1/0
C    192.168.3.0/24 is directly connected, FastEthernet0/0
```

图 5-4　路由器 R3 的路由表

步骤 7：在计算机 PC1 的命令行界面输入 ping 192.168.1.1，检验联通性，如图 5-5 所示。

```
C:\>ping 192.168.1.1

正在 Ping 192.168.1.1 具有 32 字节的数据:
来自 192.168.1.1 的回复: 字节=32 时间=10ms TTL=126
来自 192.168.1.1 的回复: 字节=32 时间=10ms TTL=126
来自 192.168.1.1 的回复: 字节=32 时间=9ms TTL=126
来自 192.168.1.1 的回复: 字节=32 时间=10ms TTL=126

192.168.1.1 的 Ping 统计信息:
    数据包: 已发送 = 4，已接收 = 4，丢失 = 0 (0% 丢失)，
往返行程的估计时间(以毫秒为单位):
    最短 = 9ms，最长 = 10ms，平均 = 9ms
```

图 5-5　从计算机 PC1 ping 计算机 PC2

步骤 8：在计算机 PC1 的命令行界面输入 ping 192.168.3.1，检验联通性，如图 5-6 所示。

```
C:\>ping 192.168.3.1

正在 Ping 192.168.3.1 具有 32 字节的数据:
来自 192.168.3.1 的回复: 字节=32 时间=18ms TTL=125
来自 192.168.3.1 的回复: 字节=32 时间=18ms TTL=125
来自 192.168.3.1 的回复: 字节=32 时间=18ms TTL=125
来自 192.168.3.1 的回复: 字节=32 时间=18ms TTL=125

192.168.3.1 的 Ping 统计信息:
    数据包: 已发送 = 4，已接收 = 4，丢失 = 0 (0% 丢失)，
往返行程的估计时间(以毫秒为单位):
    最短 = 18ms，最长 = 18ms，平均 = 18ms
```

图 5-6　从计算机 PC1 ping 计算机 PC3

步骤9：在计算机 PC2 的命令行界面输入 ping 192.168.3.1，检验联通性，如图 5-7 所示。

```
C:\>ping 192.168.3.1

正在 Ping 192.168.3.1 具有 32 字节的数据:
来自 192.168.3.1 的回复: 字节=32 时间=9ms TTL=126
来自 192.168.3.1 的回复: 字节=32 时间=9ms TTL=126
来自 192.168.3.1 的回复: 字节=32 时间=9ms TTL=126
来自 192.168.3.1 的回复: 字节=32 时间=9ms TTL=126

192.168.3.1 的 Ping 统计信息:
    数据包: 已发送 = 4，已接收 = 4，丢失 = 0 (0% 丢失)，
往返行程的估计时间(以毫秒为单位):
    最短 = 9ms，最长 = 9ms，平均 = 9ms
```

图 5-7　从计算机 PC2 ping 计算机 PC3

通过查看路由表和检验联通性可以发现，OSPF 动态路由在路由表中是路由来源为"O"、管理距离值为 110 的路由，路由表中的路由完整，计算机 PC1、PC2 和 PC3 实现了互通。

5.4.2　多区域 OSPF 的配置

本小节网络拓扑如图 5-8 所示，路由器 R1、R2 和 R3 的接口，以及计算机 PC1 和 PC2 的 IP 地址已经配置完成，要求完成多区域 OSPF 的配置，实现计算机 PC1 和 PC2 互通。

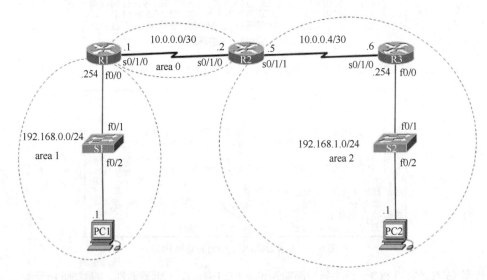

图 5-8　网络拓扑

步骤1：在路由器 R1 的全局配置模式下配置 OSPF，输入以下代码。

```
R1(config)#router ospf 1
R1(config-router)#network 10.0.0.0 0.0.0.3 area 0
R1(config-router)#network 192.168.0.0 0.0.0.255 area 1
```

步骤2：在路由器 R2 的全局配置模式下配置 OSPF，输入以下代码。

```
R2(config)#router ospf 1
R2(config-router)#network 10.0.0.0 0.0.0.3 area 0
R2(config-router)#network 10.0.0.4 0.0.0.3 area 2
```

步骤 3：在路由器 R3 的全局配置模式下配置 OSPF，输入以下代码。

```
R3(config)#router ospf 1
R3(config-router)#network 10.0.0.4 0.0.0.3 area 2
R3(config-router)#network 192.168.1.0 0.0.0.255 area 2
```

步骤 4：在路由器 R1 的特权执行模式下输入 show ip route，查看路由表，如图 5-9 所示。

```
R1#show ip route
Codes: C - connected, S - static, R - RIP, M - mobile, B - BGP
       D - EIGRP, EX - EIGRP external, O - OSPF, IA - OSPF inter area
       N1 - OSPF NSSA external type 1, N2 - OSPF NSSA external type 2
       E1 - OSPF external type 1, E2 - OSPF external type 2
       i - IS-IS, su - IS-IS summary, L1 - IS-IS level-1, L2 - IS-IS level-2
       ia - IS-IS inter area, * - candidate default, U - per-user static route
       o - ODR, P - periodic downloaded static route

Gateway of last resort is not set

     10.0.0.0/30 is subnetted, 2 subnets
C       10.0.0.0 is directly connected, Serial0/1/0
O IA    10.0.0.4 [110/1562] via 10.0.0.2, 00:01:50, Serial0/1/0
C    192.168.0.0/24 is directly connected, FastEthernet0/0
O IA 192.168.1.0/24 [110/1563] via 10.0.0.2, 00:01:50, Serial0/1/0
```

图 5-9　路由器 R1 的路由表

步骤 5：在路由器 R2 的特权执行模式下输入 show ip route，查看路由表，如图 5-10 所示。

```
R2#show ip route
Codes: C - connected, S - static, R - RIP, M - mobile, B - BGP
       D - EIGRP, EX - EIGRP external, O - OSPF, IA - OSPF inter area
       N1 - OSPF NSSA external type 1, N2 - OSPF NSSA external type 2
       E1 - OSPF external type 1, E2 - OSPF external type 2
       i - IS-IS, su - IS-IS summary, L1 - IS-IS level-1, L2 - IS-IS level-2
       ia - IS-IS inter area, * - candidate default, U - per-user static route
       o - ODR, P - periodic downloaded static route

Gateway of last resort is not set

     10.0.0.0/30 is subnetted, 2 subnets
C       10.0.0.0 is directly connected, Serial0/1/0
C       10.0.0.4 is directly connected, Serial0/1/1
O IA 192.168.0.0/24 [110/782] via 10.0.0.1, 00:01:05, Serial0/1/0
O    192.168.1.0/24 [110/782] via 10.0.0.6, 00:02:02, Serial0/1/1
```

图 5-10　路由器 R2 的路由表

步骤 6：在路由器 R3 的特权执行模式下输入 show ip route，查看路由表，如图 5-11 所示。

```
R3#show ip route
Codes: C - connected, S - static, R - RIP, M - mobile, B - BGP
       D - EIGRP, EX - EIGRP external, O - OSPF, IA - OSPF inter area
       N1 - OSPF NSSA external type 1, N2 - OSPF NSSA external type 2
       E1 - OSPF external type 1, E2 - OSPF external type 2
       i - IS-IS, su - IS-IS summary, L1 - IS-IS level-1, L2 - IS-IS level-2
       ia - IS-IS inter area, * - candidate default, U - per-user static route
       o - ODR, P - periodic downloaded static route

Gateway of last resort is not set

     10.0.0.0/30 is subnetted, 2 subnets
O IA    10.0.0.0 [110/1562] via 10.0.0.5, 00:03:19, Serial0/1/0
C       10.0.0.4 is directly connected, Serial0/1/0
O IA 192.168.0.0/24 [110/1563] via 10.0.0.5, 00:02:23, Serial0/1/0
C    192.168.1.0/24 is directly connected, FastEthernet0/0
```

图 5-11　路由器 R3 的路由表

步骤 7：在计算机 PC1 的命令行界面输入 ping 192.168.1.1，检验联通性，如图 5-12 所示。

```
C:\>ping 192.168.1.1
正在 Ping 192.168.1.1 具有 32 字节的数据：
来自 192.168.1.1 的回复：字节=32 时间=18ms TTL=125
来自 192.168.1.1 的回复：字节=32 时间=18ms TTL=125
来自 192.168.1.1 的回复：字节=32 时间=18ms TTL=125
来自 192.168.1.1 的回复：字节=32 时间=18ms TTL=125

192.168.1.1 的 Ping 统计信息：
    数据包：已发送 = 4，已接收 = 4，丢失 = 0 (0% 丢失)，
往返行程的估计时间(以毫秒为单位)：
    最短 = 18ms，最长 = 18ms，平均 = 18ms
```

图 5-12　从计算机 PC1 ping 计算机 PC2

通过查看路由表和检验联通性可以发现，当配置了多区域 OSPF 后，路由器 R1、R2 和 R3 的路由表中有路由来源为"O IA"的路由，"O IA"代表 OSPF 生成的自治系统内的区域间路由，3 台路由器的路由表中的路由完整，计算机 PC1 和 PC2 实现了互通。

5.4.3　骨干区域未与非骨干区域直连

本小节网络拓扑如图 5-13 所示，路由器 R1、R2 和 R3 的接口，以及计算机 PC1 和 PC2 的 IP 地址已经配置完成，要求完成多区域 OSPF 的配置使骨干区域与非骨干区域直连，实现计算机 PC1 和 PC2 互通，并查看、分析路由器 R1、R2 和 R3 的 Router ID。

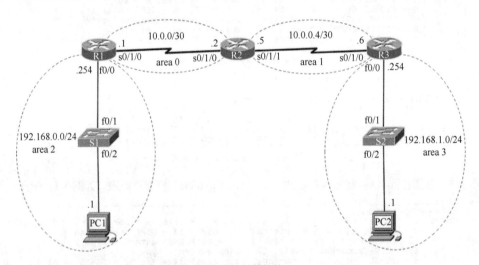

图 5-13　网络拓扑

步骤 1：在路由器 R1 的全局配置模式下配置 OSPF，输入以下代码。

```
R1(config)#router ospf 1
R1(config-router)#network 10.0.0.0 0.0.0.3 area 0
R1(config-router)#network 192.168.0.0 0.0.0.255 area 2
```

步骤 2：在路由器 R2 的全局配置模式下配置 OSPF，输入以下代码。

```
R2(config)#router ospf 1
R2(config-router)#network 10.0.0.0 0.0.0.3 area 0
R2(config-router)#network 10.0.0.4 0.0.0.3 area 1
```

步骤 3：在路由器 R3 的全局配置模式下配置 OSPF，输入以下代码。

```
R3(config)#router ospf 1
R3(config-router)#network 10.0.0.4 0.0.0.3 area 1
R3(config-router)#network 192.168.1.0 0.0.0.255 area 3
```

步骤 4：在路由器 R1 的特权执行模式下输入 show ip route，查看路由表，如图 5-14 所示。

```
R1#show ip route
Codes: C - connected, S - static, R - RIP, M - mobile, B - BGP
       D - EIGRP, EX - EIGRP external, O - OSPF, IA - OSPF inter area
       N1 - OSPF NSSA external type 1, N2 - OSPF NSSA external type 2
       E1 - OSPF external type 1, E2 - OSPF external type 2
       i - IS-IS, su - IS-IS summary, L1 - IS-IS level-1, L2 - IS-IS level-2
       ia - IS-IS inter area, * - candidate default, U - per-user static route
       o - ODR, P - periodic downloaded static route

Gateway of last resort is not set

     10.0.0.0/30 is subnetted, 2 subnets
C       10.0.0.0 is directly connected, Serial0/1/0
O IA    10.0.0.4 [110/1562] via 10.0.0.2, 01:45:16, Serial0/1/0
C    192.168.0.0/24 is directly connected, FastEthernet0/0
```

图 5-14　路由器 R1 的路由表

步骤 5：在路由器 R2 的特权执行模式下输入 show ip route，查看路由表，如图 5-15 所示。

```
R2#show ip route
Codes: C - connected, S - static, R - RIP, M - mobile, B - BGP
       D - EIGRP, EX - EIGRP external, O - OSPF, IA - OSPF inter area
       N1 - OSPF NSSA external type 1, N2 - OSPF NSSA external type 2
       E1 - OSPF external type 1, E2 - OSPF external type 2
       i - IS-IS, su - IS-IS summary, L1 - IS-IS level-1, L2 - IS-IS level-2
       ia - IS-IS inter area, * - candidate default, U - per-user static route
       o - ODR, P - periodic downloaded static route

Gateway of last resort is not set

     10.0.0.0/30 is subnetted, 2 subnets
C       10.0.0.0 is directly connected, Serial0/1/0
C       10.0.0.4 is directly connected, Serial0/1/1
O IA 192.168.0.0/24 [110/782] via 10.0.0.1, 01:44:40, Serial0/1/0
```

图 5-15　路由器 R2 的路由表

步骤 6：在路由器 R3 的特权执行模式下输入 show ip route，查看路由表，如图 5-16 所示。

```
R3#show ip route
Codes: C - connected, S - static, R - RIP, M - mobile, B - BGP
       D - EIGRP, EX - EIGRP external, O - OSPF, IA - OSPF inter area
       N1 - OSPF NSSA external type 1, N2 - OSPF NSSA external type 2
       E1 - OSPF external type 1, E2 - OSPF external type 2
       i - IS-IS, su - IS-IS summary, L1 - IS-IS level-1, L2 - IS-IS level-2
       ia - IS-IS inter area, * - candidate default, U - per-user static route
       o - ODR, P - periodic downloaded static route

Gateway of last resort is not set

     10.0.0.0/30 is subnetted, 2 subnets
O IA    10.0.0.0 [110/1562] via 10.0.0.5, 01:43:47, Serial0/1/0
C       10.0.0.4 is directly connected, Serial0/1/0
O IA 192.168.0.0/24 [110/1563] via 10.0.0.5, 01:43:47, Serial0/1/0
C    192.168.1.0/24 is directly connected, FastEthernet0/0
```

图 5-16　路由器 R3 的路由表

步骤 7：在计算机 PC1 的命令行界面输入 ping 192.168.1.1，检验联通性，如图 5-17 所示。

```
C:\>ping 192.168.1.1
正在 Ping 192.168.1.1 具有 32 字节的数据:
来自 192.168.0.254 的回复: 无法访问目标主机。
来自 192.168.0.254 的回复: 无法访问目标主机。
来自 192.168.0.254 的回复: 无法访问目标主机。
来自 192.168.0.254 的回复: 无法访问目标主机。

192.168.1.1 的 Ping 统计信息:
    数据包: 已发送 = 4, 已接收 = 4, 丢失 = 0 (0% 丢失),
```

图 5-17　从计算机 PC1 ping 计算机 PC2

通过查看路由表和检验联通性可以发现,路由器 R1 和 R2 无法获取 192.168.1.0/24(area 3)的路由,网络无法实现互通。原因是,非骨干区域(area 3)没有与骨干区域(area 0)直连。

步骤 8:在路由器 R1 的特权执行模式下输入 show ip protocols,查看路由协议的详细信息,如图 5-18 所示。

```
R1#show ip protocols
Routing Protocol is "ospf 1"
  Outgoing update filter list for all interfaces is not set
  Incoming update filter list for all interfaces is not set
  Router ID 192.168.0.254
  It is an area border router
  Number of areas in this router is 2. 2 normal 0 stub 0 nssa
  Maximum path: 4
  Routing for Networks:
    10.0.0.0 0.0.0.3 area 0
    192.168.0.0 0.0.0.255 area 2
  Reference bandwidth unit is 100 mbps
  Routing Information Sources:
    Gateway         Distance       Last Update
    10.0.0.5           110         00:03:39
  Distance: (default is 110)
```

图 5-18　查看路由器 R1 路由协议的详细信息

步骤 9:在路由器 R2 的特权执行模式下输入 show ip protocols,查看路由协议的详细信息,如图 5-19 所示。

```
R2#show ip proTOcols
Routing Protocol is "ospf 1"
  Outgoing update filter list for all interfaces is not set
  Incoming update filter list for all interfaces is not set
  Router ID 10.0.0.5
  It is an area border router
  Number of areas in this router is 2. 2 normal 0 stub 0 nssa
  Maximum path: 4
  Routing for Networks:
    10.0.0.0 0.0.0.3 area 0
    10.0.0.4 0.0.0.3 area 1
  Reference bandwidth unit is 100 mbps
  Routing Information Sources:
    Gateway         Distance       Last Update
    192.168.0.254      110         00:11:03
  Distance: (default is 110)
```

图 5-19　查看路由器 R2 路由协议的详细信息

步骤 10:在路由器 R3 的特权执行模式下输入 show ip protocols,查看路由协议的详细信息,如图 5-20 所示。

```
R3#show ip protocols
Routing Protocol is "ospf 1"
  Outgoing update filter list for all interfaces is not set
  Incoming update filter list for all interfaces is not set
  Router ID 192.168.1.254
  Number of areas in this router is 2. 2 normal 0 stub 0 nssa
  Maximum path: 4
  Routing for Networks:
    10.0.0.4 0.0.0.3 area 1
    192.168.1.0 0.0.0.255 area 3
  Reference bandwidth unit is 100 mbps
  Routing Information Sources:
    Gateway          Distance        Last Update
    10.0.0.5         110             00:11:56
  Distance: (default is 110)
```

图 5-20　查看路由器 R3 路由协议的详细信息

通过查看路由器 R1、R2 和 R3 路由协议的详细信息可以发现，路由器 R1 的 Router ID 是 192.168.0.254，路由器 R2 的 Router ID 是 10.0.0.5，路由器 R3 的 Router ID 是 192.168.1.254。OSPF Router ID 的确定原则：如果用 router-id 命令为路由器配置了 IP 地址，那么这个 IP 地址将成为 Router ID。本小节中，3 台路由器都没有使用 router-id 命令配置 IP 地址，因此路由器会选择所有环回口的最高 IP 地址。本小节中，3 台路由器也没有配置环回口，因此路由器会选择所有活动物理接口的最高 IP 地址。路由器 R1 有两个活动物理接口，两个接口的 IP 地址分别是 192.168.0.254 和 10.0.0.1，192.168.0.254 的数据值大于 10.0.0.1 的数据值，所以 192.168.0.254 成为路由器 R1 的 Router ID。路由器 R2 有两个活动物理接口，两个接口的 IP 地址分别是 10.0.0.2 和 10.0.0.5，10.0.0.5 的数据值大于 10.0.0.2 的数据值，所以 10.0.0.5 成为路由器 R2 的 Router ID。路由器 R3 有两个活动物理接口，两个接口的 IP 地址分别是 10.0.0.6 和 192.168.1.254，192.168.1.254 的数据值大于 10.0.0.6 的数据值，所以 192.168.1.254 成为路由器 R3 的 Router ID。

5.4.4　OSPF 虚链路的配置

本小节网络拓扑如图 5-21 所示，要求完成 OSPF 虚链路的配置，实现计算机 PC1 和 PC2 互通。

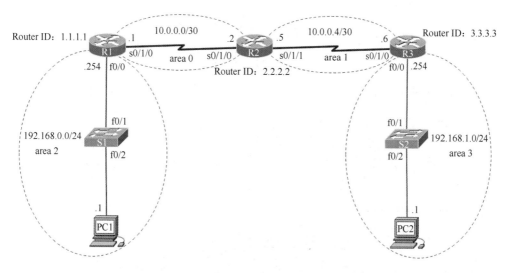

图 5-21　网络拓扑

步骤 1：在路由器 R1 的全局配置模式下配置 OSPF，输入以下代码。

```
R1(config)#router ospf 1
```

```
R1(config-router)#router-id 1.1.1.1
R1(config-router)#network 10.0.0.0 0.0.0.3 area 0
R1(config-router)#network 192.168.0.0 0.0.0.255 area 2
```

步骤 2：在路由器 R2 的全局配置模式下配置 OSPF，输入以下代码。

```
R2(config)#router ospf 1
R2(config-router)#router-id 2.2.2.2
R2(config-router)#network 10.0.0.0 0.0.0.3 area 0
R2(config-router)#network 10.0.0.4 0.0.0.3 area 1
```

步骤 3：在路由器 R3 的全局配置模式下配置 OSPF，输入以下代码。

```
R3(config)#router ospf 1
R3(config-router)#router-id 3.3.3.3
R3(config-router)#network 10.0.0.4 0.0.0.3 area 1
R3(config-router)#network 192.168.1.0 0.0.0.255 area 3
```

步骤 4：为路由器 R2 配置虚链路，输入以下代码。

```
R2(config-router)#area 1 virtual-link 3.3.3.3
```

步骤 5：为路由器 R3 配置虚链路，输入以下代码。

```
R3(config-router)#area 1 virtual-link 2.2.2.2
```

步骤 6：在路由器 R1 的特权执行模式下输入 show ip route，查看路由表，如图 5-22 所示。

```
R1#show ip route
Codes: C - connected, S - static, R - RIP, M - mobile, B - BGP
       D - EIGRP, EX - EIGRP external, O - OSPF, IA - OSPF inter area
       N1 - OSPF NSSA external type 1, N2 - OSPF NSSA external type 2
       E1 - OSPF external type 1, E2 - OSPF external type 2
       i - IS-IS, su - IS-IS summary, L1 - IS-IS level-1, L2 - IS-IS level-2
       ia - IS-IS inter area, * - candidate default, U - per-user static route
       o - ODR, P - periodic downloaded static route

Gateway of last resort is not set

     10.0.0.0/30 is subnetted, 2 subnets
C       10.0.0.0 is directly connected, Serial0/1/0
O IA    10.0.0.4 [110/1562] via 10.0.0.2, 00:05:14, Serial0/1/0
C    192.168.0.0/24 is directly connected, FastEthernet0/0
O IA 192.168.1.0/24 [110/1563] via 10.0.0.2, 00:03:24, Serial0/1/0
```

图 5-22　路由器 R1 的路由表

步骤 7：在路由器 R2 的特权执行模式下输入 show ip route，查看路由表，如图 5-23 所示。

```
R2#show ip route
Codes: C - connected, S - static, R - RIP, M - mobile, B - BGP
       D - EIGRP, EX - EIGRP external, O - OSPF, IA - OSPF inter area
       N1 - OSPF NSSA external type 1, N2 - OSPF NSSA external type 2
       E1 - OSPF external type 1, E2 - OSPF external type 2
       i - IS-IS, su - IS-IS summary, L1 - IS-IS level-1, L2 - IS-IS level-2
       ia - IS-IS inter area, * - candidate default, U - per-user static route
       o - ODR, P - periodic downloaded static route

Gateway of last resort is not set

     10.0.0.0/30 is subnetted, 2 subnets
C       10.0.0.0 is directly connected, Serial0/1/0
C       10.0.0.4 is directly connected, Serial0/1/1
O IA 192.168.0.0/24 [110/782] via 10.0.0.1, 00:03:20, Serial0/1/0
O IA 192.168.1.0/24 [110/782] via 10.0.0.6, 00:01:29, Serial0/1/1
```

图 5-23　路由器 R2 的路由表

步骤 8：在路由器 R3 的特权执行模式下输入 show ip route，查看路由表，如图 5-24 所示。

```
R3#show ip route
Codes: C - connected, S - static, R - RIP, M - mobile, B - BGP
       D - EIGRP, EX - EIGRP external, O - OSPF, IA - OSPF inter area
       N1 - OSPF NSSA external type 1, N2 - OSPF NSSA external type 2
       E1 - OSPF external type 1, E2 - OSPF external type 2
       i - IS-IS, su - IS-IS summary, L1 - IS-IS level-1, L2 - IS-IS level-2
       ia - IS-IS inter area, * - candidate default, U - per-user static route
       o - ODR, P - periodic downloaded static route

Gateway of last resort is not set

     10.0.0.0/30 is subnetted, 2 subnets
O       10.0.0.0 [110/1562] via 10.0.0.5, 00:00:59, Serial0/1/0
C       10.0.0.4 is directly connected, Serial0/1/0
O IA 192.168.0.0/24 [110/1563] via 10.0.0.5, 00:00:59, Serial0/1/0
C    192.168.1.0/24 is directly connected, FastEthernet0/0
```

图 5-24　路由器 R3 的路由表

步骤 9：在计算机 PC1 的命令行界面输入 ping 192.168.1.1，检验联通性，如图 5-25 所示。

```
C:\>ping 192.168.1.1

正在 Ping 192.168.1.1 具有 32 字节的数据：
来自 192.168.1.1 的回复：字节=32 时间=18ms TTL=125
来自 192.168.1.1 的回复：字节=32 时间=18ms TTL=125
来自 192.168.1.1 的回复：字节=32 时间=18ms TTL=125
来自 192.168.1.1 的回复：字节=32 时间=18ms TTL=125

192.168.1.1 的 Ping 统计信息：
    数据包：已发送 = 4，已接收 = 4，丢失 = 0 (0% 丢失)，
往返行程的估计时间(以毫秒为单位)：
    最短 = 18ms，最长 = 18ms，平均 = 18ms
```

图 5-25　从计算机 PC1 ping 计算机 PC2

步骤 10：在路由器 R1 的特权执行模式下输入 show ip protocols，查看路由协议的详细信息，如图 5-26 所示。

```
R1#show ip protocols
Routing Protocol is "ospf 1"
  Outgoing update filter list for all interfaces is not set
  Incoming update filter list for all interfaces is not set
  Router ID 1.1.1.1
  It is an area border router
  Number of areas in this router is 2. 2 normal 0 stub 0 nssa
  Maximum path: 4
  Routing for Networks:
    10.0.0.0 0.0.0.3 area 0
    192.168.0.0 0.0.0.255 area 2
  Reference bandwidth unit is 100 mbps
  Routing Information Sources:
    Gateway         Distance      Last Update
    3.3.3.3         110           00:04:07
    2.2.2.2         110           00:05:58
  Distance: (default is 110)
```

图 5-26　查看路由器 R1 路由协议的详细信息

步骤 11：在路由器 R2 的特权执行模式下输入 show ip protocols，查看路由协议的详细信息，如图 5-27 所示。

```
R2#show ip protocols
Routing Protocol is "ospf 1"
  Outgoing update filter list for all interfaces is not set
  Incoming update filter list for all interfaces is not set
  Router ID 2.2.2.2
  It is an area border router
  Number of areas in this router is 2. 2 normal 0 stub 0 nssa
  Maximum path: 4
  Routing for Networks:
    10.0.0.0 0.0.0.3 area 0
    10.0.0.4 0.0.0.3 area 1
  Reference bandwidth unit is 100 mbps
  Routing Information Sources:
    Gateway         Distance      Last Update
    3.3.3.3         110           00:02:14
    1.1.1.1         110           00:04:06
  Distance: (default is 110)
```

图 5-27　查看路由器 R2 路由协议的详细信息

步骤 12：在路由器 R3 的特权执行模式下输入 show ip protocols，查看路由协议的详细信息，如图 5-28 所示。

```
R3#show ip protocols
Routing Protocol is "ospf 1"
  Outgoing update filter list for all interfaces is not set
  Incoming update filter list for all interfaces is not set
  Router ID 3.3.3.3
  It is an area border router
  Number of areas in this router is 3. 3 normal 0 stub 0 nssa
  Maximum path: 4
  Routing for Networks:
    10.0.0.4 0.0.0.3 area 1
    192.168.1.0 0.0.0.255 area 3
  Reference bandwidth unit is 100 mbps
  Routing Information Sources:
    Gateway         Distance      Last Update
    1.1.1.1         110           00:05:43
    2.2.2.2         110           00:05:43
  Distance: (default is 110)
```

图 5-28　查看路由器 R3 路由协议的详细信息

本小节通过骨干区域（area 0）与非骨干区域（area 3）互联的区域（area 1）建立了一条虚链路，使非骨干区域（area 3）的路由信息能够通过虚链路通告给骨干区域。使用 router-id 命令为 3 台路由器 R1、R2 和 R3 配置了 IP 地址，通过查看路由器路由协议的详细信息可以发现，3 台路由器 R1、R2 和 R3 的 Router ID 是使用命令配置的 IP 地址。通过查看 3 台路由器的路由表和检验联通性可以发现，路由器 R1、R2 和 R3 的路由表中的路由完整，计算机 PC1 和 PC2 实现了互通。

5.4.5　不同进程 OSPF 的配置

本小节网络拓扑如图 5-29 所示，路由器 R1 和 R2 的接口，以及计算机 PC1 和 PC2 的 IP 地址已经配置完成，要求完成不同进程 OSPF 的配置，为路由器 R1 配置 OSPF 进程 1，为路由器 R2 配置 OSPF 进程 2，实现计算机 PC1 和 PC2 互通。

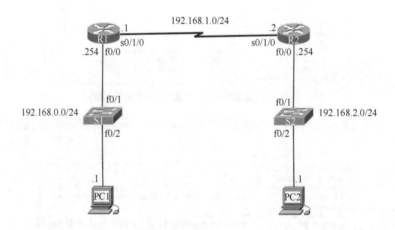

图 5-29　网络拓扑

步骤 1：在路由器 R1 的全局配置模式下配置 OSPF，输入以下代码。

```
R1(config)#router ospf 1
R1(config-router)#network 192.168.0.0 0.0.0.255 area 0
R1(config-router)#network 192.168.1.0 0.0.0.255 area 0
```

步骤 2：在路由器 R2 的全局配置模式下配置 OSPF，输入以下代码。

```
R2(config)#router ospf 2
R2(config-router)#network 192.168.1.0 0.0.0.255 area 0
R2(config-router)#network 192.168.2.0 0.0.0.255 area 0
```

步骤 3：在路由器 R1 的特权执行模式下输入 show ip route，查看路由表，如图 5-30 所示。

```
R1#show ip route
Codes: C - connected, S - static, R - RIP, M - mobile, B - BGP
       D - EIGRP, EX - EIGRP external, O - OSPF, IA - OSPF inter area
       N1 - OSPF NSSA external type 1, N2 - OSPF NSSA external type 2
       E1 - OSPF external type 1, E2 - OSPF external type 2
       i - IS-IS, su - IS-IS summary, L1 - IS-IS level-1, L2 - IS-IS level-2
       ia - IS-IS inter area, * - candidate default, U - per-user static route
       o - ODR, P - periodic downloaded static route

Gateway of last resort is not set

C    192.168.0.0/24 is directly connected, FastEthernet0/0
C    192.168.1.0/24 is directly connected, Serial0/1/0
O    192.168.2.0/24 [110/782] via 192.168.1.2, 00:00:43, Serial0/1/0
```

图 5-30　路由器 R1 的路由表

步骤 4：在路由器 R2 的特权执行模式下输入 show ip route，查看路由表，如图 5-31 所示。

```
R2#show ip route
Codes: C - connected, S - static, R - RIP, M - mobile, B - BGP
       D - EIGRP, EX - EIGRP external, O - OSPF, IA - OSPF inter area
       N1 - OSPF NSSA external type 1, N2 - OSPF NSSA external type 2
       E1 - OSPF external type 1, E2 - OSPF external type 2
       i - IS-IS, su - IS-IS summary, L1 - IS-IS level-1, L2 - IS-IS level-2
       ia - IS-IS inter area, * - candidate default, U - per-user static route
       o - ODR, P - periodic downloaded static route

Gateway of last resort is not set

O    192.168.0.0/24 [110/782] via 192.168.1.1, 00:01:31, Serial0/1/0
C    192.168.1.0/24 is directly connected, Serial0/1/0
C    192.168.2.0/24 is directly connected, FastEthernet0/0
```

图 5-31　路由器 R2 的路由表

步骤 5：在计算机 PC1 的命令行界面输入 ping 192.168.2.1，检验联通性，如图 5-32 所示。

```
C:\>ping 192.168.2.1

正在 Ping 192.168.2.1 具有 32 字节的数据:
来自 192.168.2.1 的回复: 字节=32 时间=10ms TTL=126
来自 192.168.2.1 的回复: 字节=32 时间=10ms TTL=126
来自 192.168.2.1 的回复: 字节=32 时间=10ms TTL=126
来自 192.168.2.1 的回复: 字节=32 时间=10ms TTL=126

192.168.2.1 的 Ping 统计信息:
    数据包: 已发送 = 4, 已接收 = 4, 丢失 = 0 (0% 丢失),
往返行程的估计时间(以毫秒为单位):
    最短 = 10ms, 最长 = 10ms, 平均 = 10ms
```

图 5-32　从计算机 PC1 ping 计算机 PC2

本小节为路由器 R1 配置了 OSPF 进程 1，为路由器 R2 配置了 OSPF 进程 2，通过查看路由表和检验联通性可以发现，路由器 R1 和 R2 的路由表中的路由完整，计算机 PC1 和 PC2 实现了互通。如果为不同的路由器配置不同的 OSPF 进程，因为 OSPF 的进程 ID 仅在本地有效，路由器 R1 和 R2 在建立邻接关系时不需要匹配进程 ID，所以路由器 R1 和 R2 可以建立邻接关系，两台路由器的路由表中的路由完整，网络可以实现互通。

5.5 项目小结

本项目完成了单区域 OSPF、多区域 OSPF 和 OSPF 虚链路的配置等。在区域规划过程中，当非骨干区域和骨干区域（area 0）直连时，配置多区域 OSPF 可以使网络互通；当非骨干区域没有与骨干区域（area 0）直连时，配置多区域 OSPF 无法直接使网络互通，这种情况可以配置虚链路，使未与骨干区域直连的非骨干区域通过虚链路与骨干区域交换路由信息，实现网络互通。因为 OSPF 进程 ID 只在本地有效，所以当不同的路由器分别运行不同的 OSPF 进程时，网络可以实现互通。

5.6 拓展训练

本项目的拓展训练网络拓扑如图 5-33 所示，要求完成如下配置：
（1）完成路由器接口和计算机 IP 地址的配置；
（2）完成不同进程多区域 OSPF 的配置，OSPF 进程规划和区域规划按照网络拓扑中的标注进行设置；
（3）完成 OSPF 虚链路的配置，实现计算机 PC1 和 PC2 互通；
（4）要求业务网段中不出现协议报文。

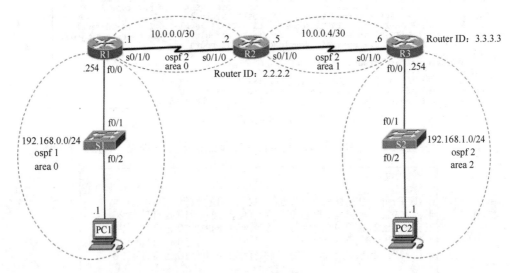

图 5-33 项目 5 拓展训练网络拓扑

项目 6
路由重分布

6.1 用户需求

某学校网络拓扑如图 6-1 所示,学校校园分为两个区:A 区和 B 区(B 区是后来扩建的)。A 区网络运行的是 RIPv2,B 区网络运行的是 OSPF,怎样实现网络的互通?

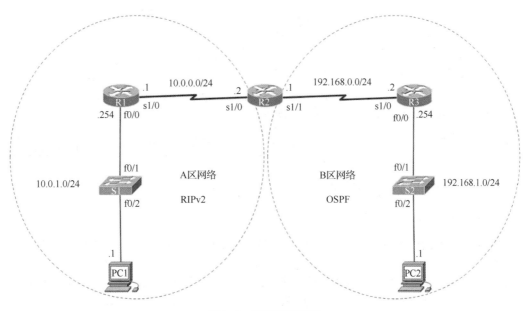

图 6-1 某学校网络拓扑

6.2 知识梳理

6.2.1 路由重分布的概念

如果网络运行多种路由协议(同一种协议的多个实例被视为多种不同的协议),要在路由协议之间交换路由信息,则需要进行路由重分布。路由重分布是指连接到不同路由域的边界路由器在路由协议之间交换和通告路由选择信息。

6.2.2 路由重分布的种类

路由重分布主要有双向重分布和单向重分布两种。
① 双向重分布：在两个路由协议之间重分布所有路由。
② 单向重分布：只将通过一种路由协议获悉的网络重分布给另一种路由协议，同时只将通过该路由协议获得的网络传递给其他路由协议。

路由重分布的种类和种子度量值

6.2.3 种子度量值

路由重分布时，路由器通告与接口相连的链路时使用的默认度量值（即种子度量值）是在重分布配置期间被定义的。常用的种子度量值如表6-1所示。

表6-1 种子度量值

将路由重分布到该协议中	种子度量值
路由信息协议（Routing Information Protocol，RIP）	无穷大
内部网关路由协议（Interior Gateway Routing Protocol，IGRP）、增强型内部网关路由协议（Enhanced Interior Gateway Routing Protocol，EIGRP）	无穷大
开放最短路径优先（Open Shortest Path First，OSPF）协议	BGP 路由的种子度量值为 1，其他路由的种子度量值为 20。在 OSPF 和 OSPF 之间进行路由重分布时，区域内路由和区域间路由的种子度量值都保持不变
中间系统到中间系统（Intermediate System to Intermediate System，IS-IS）	0
边界网关协议（Border Gateway Protocol，BGP）	BGP 的种子度量值被设置为 IGP 的种子度量值

6.2.4 路由重分布的配置

1. 重分布到 RIP

Router(config-router)#redistribute *protocol* [match *route-type*] [metric *metric-value*] [route-map *map-tag*]

protocol：路由重分布的源协议，可选的关键字有 bgp、static、connected、eigrp、isis、ospf 和 rip 等。
route-type：可选参数，将 OSPF 路由重分布到另外一种路由协议时使用的参数。
metric-value：可选参数，用于指定重分布的路由的 RIP 种子度量值。
map-tag：指定路由映射表的标识符。重分布时将查询它以便过滤从源路由协议引入 RIP 中的路由。

2. 重分布到 OSPF

Router(config-router)#redistribute *protocol* [*process-id*] [metric *metric-value*] [metric-type *type-value*][route-map *map-tag*] [subnets] [tag *tag-value*]

protocol：路由重分布的源协议，可选的关键字有 bgp、static、connected、eigrp、isis、ospf 和 rip 等。
process-id：对于 EIGRP 或者 BGP，该参数值为自治系统号；对于 OSPF，该参数值为进程 ID；对于 IS-IS，不需要该参数。
metric-value：可选参数，用于指定重分布的路由的 OSPF 种子度量值。
type-value：可选参数，指定通告到 OSPF 路由选择域的外部路由的链路类型。取值为 1 表示 1 类外部路由，取值为 2 表示 2 类外部路由，默认值为 2。

map-tag：指定路由映射表的标识符。重分布时将查询它以便过滤从源路由协议引入 OSPF 中的路由。

subnets：可选参数，指定应该同时重分布的子网路由。

tag-value：可选参数，一个 32 位的十进制数，被附加到每条外部路由上。它可在 OSPF 自治系统边界路由器（ASBR）之间交换信息。

6.3 方案设计

图 6-1 所示的网络拓扑中运行了 RIPv2 和 OSPF 两种路由协议。要实现网络互通，可以进行路由重分布的配置，把 RIP 的路由重分布到 OSPF，把 OSPF 的路由重分布到 RIP。在有些网络中，可能需要把静态路由重分布到 RIP 或者 OSPF。

6.4 项目实施

6.4.1 将静态路由重分布到 RIP

某学校的网络拓扑如图 6-2 所示，路由器 R1、交换机 S1 和计算机 PC1 位于学校内网，路由器 R2 为学校网络出口路由器，路由器 R3、交换机 S2 和计算机 PC2 位于外网。路由器 R1、R2 和 R3，以及计算机 PC1 和 PC2 的 IP 地址已经配置完成，学校内网运行 RIPv2，学校内网与外网的互通使用静态路由（不允许使用默认路由）实现，要求配置路由和路由重分布，把静态路由重分布到 RIP，实现计算机 PC1 和 PC2 互通。

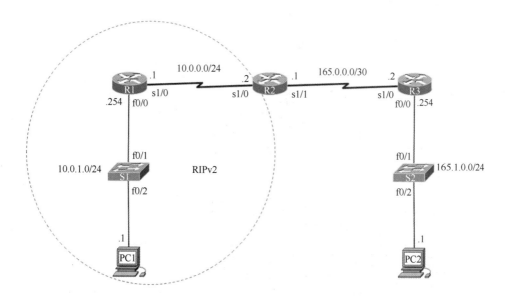

图 6-2　某学校的网络拓扑

步骤 1：在路由器 R1 的全局配置模式下配置 RIPv2，输入以下代码。

```
R1(config)#router rip
```

```
R1(config-router)#version 2
R1(config-router)#network 10.0.0.0
R1(config-router)#no auto-summary
```

步骤 2：在路由器 R2 的全局配置模式下配置 RIPv2，输入以下代码。

```
R2(config)#router rip
R2(config-router)#version 2
R2(config-router)#network 10.0.0.0
R2(config-router)#no auto-summary
R2(config-router)#ip route 165.1.0.0 255.255.255.0 165.0.0.2
```

步骤 3：在路由器 R3 的全局配置模式下配置静态路由，输入以下代码。

```
R3(config)#ip route 10.0.0.0 255.255.254.0 165.0.0.1
```

步骤 4：在路由器 R1 的特权执行模式下输入 show ip route，查看路由表，如图 6-3 所示。

```
R1#show ip route
Codes: L - local, C - connected, S - static, R - RIP, M - mobile, B - BGP
       D - EIGRP, EX - EIGRP external, O - OSPF, IA - OSPF inter area
       N1 - OSPF NSSA external type 1, N2 - OSPF NSSA external type 2
       E1 - OSPF external type 1, E2 - OSPF external type 2
       i - IS-IS, su - IS-IS summary, L1 - IS-IS level-1, L2 - IS-IS level-2
       ia - IS-IS inter area, * - candidate default, U - per-user static route
       o - ODR, P - periodic downloaded static route, H - NHRP, l - LISP
       + - replicated route, % - next hop override

Gateway of last resort is not set

      10.0.0.0/8 is variably subnetted, 4 subnets, 2 masks
C        10.0.0.0/24 is directly connected, Serial1/0
L        10.0.0.1/32 is directly connected, Serial1/0
C        10.0.1.0/24 is directly connected, FastEthernet0/0
L        10.0.1.254/32 is directly connected, FastEthernet0/0
```

图 6-3　路由器 R1 的路由表

步骤 5：在路由器 R2 的特权执行模式下输入 show ip route，查看路由表，如图 6-4 所示。

```
R2#show ip route
Codes: L - local, C - connected, S - static, R - RIP, M - mobile, B - BGP
       D - EIGRP, EX - EIGRP external, O - OSPF, IA - OSPF inter area
       N1 - OSPF NSSA external type 1, N2 - OSPF NSSA external type 2
       E1 - OSPF external type 1, E2 - OSPF external type 2
       i - IS-IS, su - IS-IS summary, L1 - IS-IS level-1, L2 - IS-IS level-2
       ia - IS-IS inter area, * - candidate default, U - per-user static route
       o - ODR, P - periodic downloaded static route, H - NHRP, l - LISP
       + - replicated route, % - next hop override

Gateway of last resort is not set

      10.0.0.0/8 is variably subnetted, 3 subnets, 2 masks
C        10.0.0.0/24 is directly connected, Serial1/0
L        10.0.0.2/32 is directly connected, Serial1/0
R        10.0.1.0/24 [120/1] via 10.0.0.1, 00:00:12, Serial1/0
      165.0.0.0/16 is variably subnetted, 2 subnets, 2 masks
C        165.0.0.0/30 is directly connected, Serial1/1
L        165.0.0.1/32 is directly connected, Serial1/1
      165.1.0.0/24 is subnetted, 1 subnets
S        165.1.0.0 [1/0] via 165.0.0.2
```

图 6-4　路由器 R2 的路由表

步骤 6：在路由器 R3 的特权执行模式下输入 show ip route，查看路由表，如图 6-5 所示。

```
R3#show ip route
Codes: L - local, C - connected, S - static, R - RIP, M - mobile, B - BGP
       D - EIGRP, EX - EIGRP external, O - OSPF, IA - OSPF inter area
       N1 - OSPF NSSA external type 1, N2 - OSPF NSSA external type 2
       E1 - OSPF external type 1, E2 - OSPF external type 2
       i - IS-IS, su - IS-IS summary, L1 - IS-IS level-1, L2 - IS-IS level-2
       ia - IS-IS inter area, * - candidate default, U - per-user static route
       o - ODR, P - periodic downloaded static route, H - NHRP, l - LISP
       + - replicated route, % - next hop override

Gateway of last resort is not set

      10.0.0.0/23 is subnetted, 1 subnets
S        10.0.0.0 [1/0] via 165.0.0.1
      165.0.0.0/16 is variably subnetted, 2 subnets, 2 masks
C        165.0.0.0/30 is directly connected, Serial1/0
L        165.0.0.2/32 is directly connected, Serial1/0
      165.1.0.0/16 is variably subnetted, 2 subnets, 2 masks
C        165.1.0.0/24 is directly connected, FastEthernet0/0
L        165.1.0.254/32 is directly connected, FastEthernet0/0
```

图 6-5 路由器 R3 的路由表

步骤 7：配置静态路由重分布，在路由器 R2 的全局配置模式下输入以下代码。

```
R2(config)#router rip
R2(config-router)#redistribute static
```

步骤 8：在路由器 R1 的特权执行模式下输入 show ip route，查看路由表，如图 6-6 所示。

```
R1#show ip route
Codes: L - local, C - connected, S - static, R - RIP, M - mobile, B - BGP
       D - EIGRP, EX - EIGRP external, O - OSPF, IA - OSPF inter area
       N1 - OSPF NSSA external type 1, N2 - OSPF NSSA external type 2
       E1 - OSPF external type 1, E2 - OSPF external type 2
       i - IS-IS, su - IS-IS summary, L1 - IS-IS level-1, L2 - IS-IS level-2
       ia - IS-IS inter area, * - candidate default, U - per-user static route
       o - ODR, P - periodic downloaded static route, H - NHRP, l - LISP
       + - replicated route, % - next hop override

Gateway of last resort is not set

      10.0.0.0/8 is variably subnetted, 4 subnets, 2 masks
C        10.0.0.0/24 is directly connected, Serial1/0
L        10.0.0.1/32 is directly connected, Serial1/0
C        10.0.1.0/24 is directly connected, FastEthernet0/0
L        10.0.1.254/32 is directly connected, FastEthernet0/0
      165.1.0.0/24 is subnetted, 1 subnets
R        165.1.0.0 [120/1] via 10.0.0.2, 00:00:06, Serial1/0
```

图 6-6 路由器 R1 的路由表

步骤 9：在路由器 R2 的特权执行模式下输入 show ip route，查看路由表，如图 6-7 所示。

```
R2#show ip route
Codes: L - local, C - connected, S - static, R - RIP, M - mobile, B - BGP
       D - EIGRP, EX - EIGRP external, O - OSPF, IA - OSPF inter area
       N1 - OSPF NSSA external type 1, N2 - OSPF NSSA external type 2
       E1 - OSPF external type 1, E2 - OSPF external type 2
       i - IS-IS, su - IS-IS summary, L1 - IS-IS level-1, L2 - IS-IS level-2
       ia - IS-IS inter area, * - candidate default, U - per-user static route
       o - ODR, P - periodic downloaded static route, H - NHRP, l - LISP
       + - replicated route, % - next hop override

Gateway of last resort is not set

      10.0.0.0/8 is variably subnetted, 3 subnets, 2 masks
C        10.0.0.0/24 is directly connected, Serial1/0
L        10.0.0.2/32 is directly connected, Serial1/0
R        10.0.1.0/24 [120/1] via 10.0.0.1, 00:00:08, Serial1/0
      165.0.0.0/16 is variably subnetted, 2 subnets, 2 masks
C        165.0.0.0/30 is directly connected, Serial1/1
L        165.0.0.1/32 is directly connected, Serial1/1
      165.1.0.0/24 is subnetted, 1 subnets
S        165.1.0.0 [1/0] via 165.0.0.2
```

图 6-7 路由器 R2 的路由表

步骤10：在计算机 PC1 的命令行界面输入 ping 165.1.0.1，检验联通性，如图 6-8 所示。

```
C:\>ping 165.1.0.1
正在 Ping 165.1.0.1 具有 32 字节的数据:
来自 165.1.0.1 的回复: 字节=32 时间=19ms TTL=125
来自 165.1.0.1 的回复: 字节=32 时间=18ms TTL=125
来自 165.1.0.1 的回复: 字节=32 时间=19ms TTL=125
来自 165.1.0.1 的回复: 字节=32 时间=19ms TTL=125

165.1.0.1 的 Ping 统计信息:
    数据包: 已发送 = 4, 已接收 = 4, 丢失 = 0 (0% 丢失),
往返行程的估计时间(以毫秒为单位):
    最短 = 18ms, 最长 = 19ms, 平均 = 18ms
```

图 6-8 从计算机 PC1 ping 计算机 PC2

通过查看路由表可以发现，将静态路由重分布到 RIP 后，RIP 路由域内的路由器（边界路由器 R2 除外）的路由表中会出现路由来源为"R"、目的网络为 165.1.0.0 的路由，这就是重分布到 RIP 的静态路由。通过检验联通性可以发现，计算机 PC1 和 PC2 实现了互通。

6.4.2　将静态路由重分布到 OSPF

某学校的网络拓扑如图 6-9 所示，路由器 R1、交换机 S1 和计算机 PC1 位于学校内网，路由器 R2 为学校内网出口路由器，路由器 R3、交换机 S2 和计算机 PC2 位于外网。路由器 R1、R2 和 R3，以及计算机 PC1 和 PC2 的 IP 地址已经配置完成，学校内网运行 OSPF，学校内网与外网的互通使用静态路由（不允许使用默认路由）实现，要求配置路由和路由重分布，把静态路由重分布到 OSPF，实现计算机 PC1 和 PC2 互通。

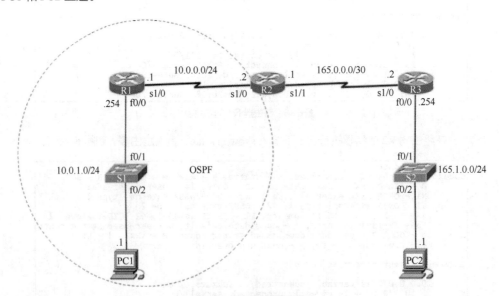

图 6-9 某学校网络拓扑

步骤1：在路由器 R1 的全局配置模式下配置 OSPF，输入以下代码。

```
R1(config)#router ospf 1
R1(config-router)#network 10.0.0.0 0.0.0.255 area 0
R1(config-router)#network 10.0.1.0 0.0.0.255 area 0
```

步骤 2：在路由器 R2 的全局配置模式下配置 OSPF，输入以下代码。

```
R2(config)#router ospf 1
R2(config-router)#network 10.0.0.0 0.0.0.255 area 0
R2(config-router)#ip route 165.1.0.0 255.255.255.0 165.0.0.2
```

步骤 3：在路由器 R3 的全局配置模式下配置静态路由，输入以下代码。

```
R3(config)#ip route 10.0.0.0 255.255.254.0 165.0.0.1
```

步骤 4：在路由器 R1 的特权执行模式下输入 show ip route，查看路由表，如图 6-10 所示。

```
R1#show ip route
Codes: L - local, C - connected, S - static, R - RIP, M - mobile, B - BGP
       D - EIGRP, EX - EIGRP external, O - OSPF, IA - OSPF inter area
       N1 - OSPF NSSA external type 1, N2 - OSPF NSSA external type 2
       E1 - OSPF external type 1, E2 - OSPF external type 2
       i - IS-IS, su - IS-IS summary, L1 - IS-IS level-1, L2 - IS-IS level-2
       ia - IS-IS inter area, * - candidate default, U - per-user static route
       o - ODR, P - periodic downloaded static route, H - NHRP, l - LISP
       + - replicated route, % - next hop override

Gateway of last resort is not set

      10.0.0.0/8 is variably subnetted, 4 subnets, 2 masks
C        10.0.0.0/24 is directly connected, Serial1/0
L        10.0.0.1/32 is directly connected, Serial1/0
C        10.0.1.0/24 is directly connected, FastEthernet0/0
L        10.0.1.254/32 is directly connected, FastEthernet0/0
```

图 6-10　路由器 R1 的路由表

步骤 5：在路由器 R2 的特权执行模式下输入 show ip route，查看路由表，如图 6-11 所示。

```
R2#show ip route
Codes: L - local, C - connected, S - static, R - RIP, M - mobile, B - BGP
       D - EIGRP, EX - EIGRP external, O - OSPF, IA - OSPF inter area
       N1 - OSPF NSSA external type 1, N2 - OSPF NSSA external type 2
       E1 - OSPF external type 1, E2 - OSPF external type 2
       i - IS-IS, su - IS-IS summary, L1 - IS-IS level-1, L2 - IS-IS level-2
       ia - IS-IS inter area, * - candidate default, U - per-user static route
       o - ODR, P - periodic downloaded static route, H - NHRP, l - LISP
       + - replicated route, % - next hop override

Gateway of last resort is not set

      10.0.0.0/8 is variably subnetted, 3 subnets, 2 masks
C        10.0.0.0/24 is directly connected, Serial1/0
L        10.0.0.2/32 is directly connected, Serial1/0
O        10.0.1.0/24 [110/65] via 10.0.0.1, 00:00:50, Serial1/0
      165.0.0.0/16 is variably subnetted, 2 subnets, 2 masks
C        165.0.0.0/30 is directly connected, Serial1/1
L        165.0.0.1/32 is directly connected, Serial1/1
      165.1.0.0/24 is subnetted, 1 subnets
S        165.1.0.0 [1/0] via 165.0.0.2
```

图 6-11　路由器 R2 的路由表

步骤 6：在路由器 R3 的特权执行模式下输入 show ip route，查看路由表，如图 6-12 所示。

```
R3#show ip route
Codes: L - local, C - connected, S - static, R - RIP, M - mobile, B - BGP
       D - EIGRP, EX - EIGRP external, O - OSPF, IA - OSPF inter area
       N1 - OSPF NSSA external type 1, N2 - OSPF NSSA external type 2
       E1 - OSPF external type 1, E2 - OSPF external type 2
       i - IS-IS, su - IS-IS summary, L1 - IS-IS level-1, L2 - IS-IS level-2
       ia - IS-IS inter area, * - candidate default, U - per-user static route
       o - ODR, P - periodic downloaded static route, H - NHRP, l - LISP
       + - replicated route, % - next hop override

Gateway of last resort is not set

      10.0.0.0/23 is subnetted, 1 subnets
S        10.0.0.0 [1/0] via 165.0.0.1
      165.0.0.0/16 is variably subnetted, 2 subnets, 2 masks
C        165.0.0.0/30 is directly connected, Serial1/0
L        165.0.0.2/32 is directly connected, Serial1/0
      165.1.0.0/16 is variably subnetted, 2 subnets, 2 masks
C        165.1.0.0/24 is directly connected, FastEthernet0/0
L        165.1.0.254/32 is directly connected, FastEthernet0/0
```

图 6-12　路由器 R3 的路由表

步骤 7：将静态路由重分布到 OSPF，在路由器 R2 的全局配置模式下输入以下代码。

```
R2(config)#router ospf 1
R2(config-router)#redistribute static subnets
```

步骤 8：在路由器 R1 的特权执行模式下输入 show ip route，查看路由表，如图 6-13 所示。

```
R1#show ip route
Codes: L - local, C - connected, S - static, R - RIP, M - mobile, B - BGP
       D - EIGRP, EX - EIGRP external, O - OSPF, IA - OSPF inter area
       N1 - OSPF NSSA external type 1, N2 - OSPF NSSA external type 2
       E1 - OSPF external type 1, E2 - OSPF external type 2
       i - IS-IS, su - IS-IS summary, L1 - IS-IS level-1, L2 - IS-IS level-2
       ia - IS-IS inter area, * - candidate default, U - per-user static route
       o - ODR, P - periodic downloaded static route, H - NHRP, l - LISP
       + - replicated route, % - next hop override

Gateway of last resort is not set

      10.0.0.0/8 is variably subnetted, 4 subnets, 2 masks
C        10.0.0.0/24 is directly connected, Serial1/0
L        10.0.0.1/32 is directly connected, Serial1/0
C        10.0.1.0/24 is directly connected, FastEthernet0/0
L        10.0.1.254/32 is directly connected, FastEthernet0/0
      165.1.0.0/24 is subnetted, 1 subnets
O E2     165.1.0.0 [110/20] via 10.0.0.2, 00:00:07, Serial1/0
```

图 6-13　路由器 R1 的路由表

步骤 9：在路由器 R2 的特权执行模式下输入 show ip route，查看路由表，如图 6-14 所示。

```
R2#show ip route
Codes: L - local, C - connected, S - static, R - RIP, M - mobile, B - BGP
       D - EIGRP, EX - EIGRP external, O - OSPF, IA - OSPF inter area
       N1 - OSPF NSSA external type 1, N2 - OSPF NSSA external type 2
       E1 - OSPF external type 1, E2 - OSPF external type 2
       i - IS-IS, su - IS-IS summary, L1 - IS-IS level-1, L2 - IS-IS level-2
       ia - IS-IS inter area, * - candidate default, U - per-user static route
       o - ODR, P - periodic downloaded static route, H - NHRP, l - LISP
       + - replicated route, % - next hop override

Gateway of last resort is not set

      10.0.0.0/8 is variably subnetted, 3 subnets, 2 masks
C        10.0.0.0/24 is directly connected, Serial1/0
L        10.0.0.2/32 is directly connected, Serial1/0
O        10.0.1.0/24 [110/65] via 10.0.0.1, 00:03:52, Serial1/0
      165.0.0.0/16 is variably subnetted, 2 subnets, 2 masks
C        165.0.0.0/30 is directly connected, Serial1/1
L        165.0.0.1/32 is directly connected, Serial1/1
      165.1.0.0/24 is subnetted, 1 subnets
S        165.1.0.0 [1/0] via 165.0.0.2
```

图 6-14　路由器 R2 的路由表

步骤 10：在计算机 PC1 的命令行界面输入 ping 165.1.0.1，检验联通性，如图 6-15 所示。

```
C:\>ping 165.1.0.1

正在 Ping 165.1.0.1 具有 32 字节的数据：
来自 165.1.0.1 的回复：字节=32 时间=18ms TTL=125
来自 165.1.0.1 的回复：字节=32 时间=18ms TTL=125
来自 165.1.0.1 的回复：字节=32 时间=18ms TTL=125
来自 165.1.0.1 的回复：字节=32 时间=18ms TTL=125

165.1.0.1 的 Ping 统计信息：
    数据包：已发送 = 4，已接收 = 4，丢失 = 0 (0% 丢失)，
往返行程的估计时间(以毫秒为单位)：
    最短 = 18ms，最长 = 18ms，平均 = 18ms
```

图 6-15　从计算机 PC1 ping 计算机 PC2

通过查看路由表可以发现，将静态路由重分布到 OSPF 后，OSPF 路由域内的路由器（边界路由器 R2 除外）的路由表中会出现路由来源为"O E2"、目的地址为 165.1.0.0 的路由，这就是重分布到 OSPF 的静态路由，"O E2"代表自治系统外的 2 类外部路由。通过检验联通性可以发现，计算机 PC1 和 PC2 实现了互通。

6.4.3 RIP 和 OSPF 路由重分布的配置

本小节网络拓扑如图 6-1 所示，路由器 R1、R2 和 R3 的接口，以及计算机 PC1 和 PC2 的 IP 地址已经配置完成，要求完成 RIP、OSPF 及 RIP 和 OSPF 路由重分布的配置，实现计算机 PC1 和 PC2 互通。

步骤 1：在路由器 R1 的全局配置模式下配置 RIPv2，输入以下代码。

```
R1(config)#router rip
R1(config-router)#version 2
R1(config-router)#network 10.0.0.0
R1(config-router)#no auto-summary
```

步骤 2：在路由器 R2 的全局配置模式下配置 RIPv2，输入以下代码。

```
R2(config)#router rip
R2(config-router)#version 2
R2(config-router)#network 10.0.0.0
R2(config-router)#no auto-summary
R2(config-router)#router ospf 1
R2(config-router)#network 192.168.0.0 0.0.0.255 area 0
```

步骤 3：在路由器 R3 的全局配置模式下配置 OSPF，输入以下代码。

```
R3(config)#router ospf 1
R3(config-router)#network 192.168.0.0 0.0.0.255 area 0
R3(config-router)#network 192.168.1.0 0.0.0.255 area 0
```

步骤 4：在路由器 R1 的特权执行模式下输入 show ip route，查看路由表，如图 6-16 所示。

```
R1#show ip route
Codes: L - local, C - connected, S - static, R - RIP, M - mobile, B - BGP
       D - EIGRP, EX - EIGRP external, O - OSPF, IA - OSPF inter area
       N1 - OSPF NSSA external type 1, N2 - OSPF NSSA external type 2
       E1 - OSPF external type 1, E2 - OSPF external type 2
       i - IS-IS, su - IS-IS summary, L1 - IS-IS level-1, L2 - IS-IS level-2
       ia - IS-IS inter area, * - candidate default, U - per-user static route
       o - ODR, P - periodic downloaded static route, H - NHRP, l - LISP
       + - replicated route, % - next hop override

Gateway of last resort is not set

      10.0.0.0/8 is variably subnetted, 4 subnets, 2 masks
C        10.0.0.0/24 is directly connected, Serial1/0
L        10.0.0.1/32 is directly connected, Serial1/0
C        10.0.1.0/24 is directly connected, FastEthernet0/0
L        10.0.1.254/32 is directly connected, FastEthernet0/0
```

图 6-16　路由器 R1 的路由表

步骤 5：在路由器 R2 的特权执行模式下输入 show ip route，查看路由表，如图 6-17 所示。

```
R2#show ip route
Codes: L - local, C - connected, S - static, R - RIP, M - mobile, B - BGP
       D - EIGRP, EX - EIGRP external, O - OSPF, IA - OSPF inter area
       N1 - OSPF NSSA external type 1, N2 - OSPF NSSA external type 2
       E1 - OSPF external type 1, E2 - OSPF external type 2
       i - IS-IS, su - IS-IS summary, L1 - IS-IS level-1, L2 - IS-IS level-2
       ia - IS-IS inter area, * - candidate default, U - per-user static route
       o - ODR, P - periodic downloaded static route, H - NHRP, l - LISP
       + - replicated route, % - next hop override

Gateway of last resort is not set

      10.0.0.0/8 is variably subnetted, 3 subnets, 2 masks
C        10.0.0.0/24 is directly connected, Serial1/0
L        10.0.0.2/32 is directly connected, Serial1/0
R        10.0.1.0/24 [120/1] via 10.0.0.1, 00:00:03, Serial1/0
      192.168.0.0/24 is variably subnetted, 2 subnets, 2 masks
C        192.168.0.0/24 is directly connected, Serial1/1
L        192.168.0.1/32 is directly connected, Serial1/1
O        192.168.1.0/24 [110/65] via 192.168.0.2, 00:02:50, Serial1/1
```

图 6-17 路由器 R2 的路由表

步骤 6：在路由器 R3 的特权执行模式下输入 show ip route，查看路由表，如图 6-18 所示。

```
R3#show ip route
Codes: L - local, C - connected, S - static, R - RIP, M - mobile, B - BGP
       D - EIGRP, EX - EIGRP external, O - OSPF, IA - OSPF inter area
       N1 - OSPF NSSA external type 1, N2 - OSPF NSSA external type 2
       E1 - OSPF external type 1, E2 - OSPF external type 2
       i - IS-IS, su - IS-IS summary, L1 - IS-IS level-1, L2 - IS-IS level-2
       ia - IS-IS inter area, * - candidate default, U - per-user static route
       o - ODR, P - periodic downloaded static route, H - NHRP, l - LISP
       + - replicated route, % - next hop override

Gateway of last resort is not set

      192.168.0.0/24 is variably subnetted, 2 subnets, 2 masks
C        192.168.0.0/24 is directly connected, Serial1/0
L        192.168.0.2/32 is directly connected, Serial1/0
      192.168.1.0/24 is variably subnetted, 2 subnets, 2 masks
C        192.168.1.0/24 is directly connected, FastEthernet0/0
L        192.168.1.254/32 is directly connected, FastEthernet0/0
```

图 6-18 路由器 R3 的路由表

步骤 7：把 OSPF 路由重分布到 RIP，在路由器 R2 的全局配置模式下输入以下代码。

```
R2(config)#router rip
R2(config-router)#redistribute ospf 1 metric 2
```

把路由重分布到 RIP 的种子度量值是无穷大的，RIP 度量值超过 15 就认为不可达。如果把 OSPF 路由重分布到 RIP，使用种子度量值（不修改度量值），那么 RIP 会认为该路由不可达，路由表中不会出现对应的路由。所以把 OSPF 路由重分布到 RIP 需要修改度量值，本小节把度量值改为 2。

步骤 8：在路由器 R1 的特权执行模式下输入 show ip route，查看路由表，如图 6-19 所示。

```
R1#show ip route
Codes: L - local, C - connected, S - static, R - RIP, M - mobile, B - BGP
       D - EIGRP, EX - EIGRP external, O - OSPF, IA - OSPF inter area
       N1 - OSPF NSSA external type 1, N2 - OSPF NSSA external type 2
       E1 - OSPF external type 1, E2 - OSPF external type 2
       i - IS-IS, su - IS-IS summary, L1 - IS-IS level-1, L2 - IS-IS level-2
       ia - IS-IS inter area, * - candidate default, U - per-user static route
       o - ODR, P - periodic downloaded static route, H - NHRP, l - LISP
       + - replicated route, % - next hop override

Gateway of last resort is not set

      10.0.0.0/8 is variably subnetted, 4 subnets, 2 masks
C        10.0.0.0/24 is directly connected, Serial1/0
L        10.0.0.1/32 is directly connected, Serial1/0
C        10.0.1.0/24 is directly connected, FastEthernet0/0
L        10.0.1.254/32 is directly connected, FastEthernet0/0
R     192.168.0.0/24 [120/2] via 10.0.0.2, 00:00:25, Serial1/0
R     192.168.1.0/24 [120/2] via 10.0.0.2, 00:00:25, Serial1/0
```

图 6-19 路由器 R1 的路由表

步骤 9：把 RIP 路由重分布到 OSPF，在路由器 R2 的全局配置模式下输入以下代码。

```
R2(config)#router ospf 1
R2(config-router)#redistribute rip subnets
```

步骤 10：在路由器 R3 的特权执行模式下输入 show ip route，查看路由表，如图 6-20 所示。

```
R3#show ip route
Codes: L - local, C - connected, S - static, R - RIP, M - mobile, B - BGP
       D - EIGRP, EX - EIGRP external, O - OSPF, IA - OSPF inter area
       N1 - OSPF NSSA external type 1, N2 - OSPF NSSA external type 2
       E1 - OSPF external type 1, E2 - OSPF external type 2
       i - IS-IS, su - IS-IS summary, L1 - IS-IS level-1, L2 - IS-IS level-2
       ia - IS-IS inter area, * - candidate default, U - per-user static route
       o - ODR, P - periodic downloaded static route, H - NHRP, l - LISP
       + - replicated route, % - next hop override

Gateway of last resort is not set

      10.0.0.0/24 is subnetted, 2 subnets
O E2     10.0.0.0 [110/20] via 192.168.0.1, 00:01:33, Serial1/0
O E2     10.0.1.0 [110/20] via 192.168.0.1, 00:01:33, Serial1/0
      192.168.0.0/24 is variably subnetted, 2 subnets, 2 masks
C        192.168.0.0/24 is directly connected, Serial1/0
L        192.168.0.2/32 is directly connected, Serial1/0
      192.168.1.0/24 is variably subnetted, 2 subnets, 2 masks
C        192.168.1.0/24 is directly connected, FastEthernet0/0
L        192.168.1.254/32 is directly connected, FastEthernet0/0
```

图 6-20　路由器 R3 的路由表

步骤 11：在路由器 R2 的特权执行模式下输入 show ip route，查看路由表，如图 6-21 所示。

```
R2#show ip route
Codes: L - local, C - connected, S - static, R - RIP, M - mobile, B - BGP
       D - EIGRP, EX - EIGRP external, O - OSPF, IA - OSPF inter area
       N1 - OSPF NSSA external type 1, N2 - OSPF NSSA external type 2
       E1 - OSPF external type 1, E2 - OSPF external type 2
       i - IS-IS, su - IS-IS summary, L1 - IS-IS level-1, L2 - IS-IS level-2
       ia - IS-IS inter area, * - candidate default, U - per-user static route
       o - ODR, P - periodic downloaded static route, H - NHRP, l - LISP
       + - replicated route, % - next hop override

Gateway of last resort is not set

      10.0.0.0/8 is variably subnetted, 3 subnets, 2 masks
C        10.0.0.0/24 is directly connected, Serial1/0
L        10.0.0.2/32 is directly connected, Serial1/0
R        10.0.1.0/24 [120/1] via 10.0.0.1, 00:00:03, Serial1/0
      192.168.0.0/24 is variably subnetted, 2 subnets, 2 masks
C        192.168.0.0/24 is directly connected, Serial1/1
L        192.168.0.1/32 is directly connected, Serial1/1
O     192.168.1.0/24 [110/65] via 192.168.0.2, 00:02:50, Serial1/1
```

图 6-21　路由器 R2 的路由表

步骤 12：在计算机 PC1 的命令行界面输入 ping 192.168.1.1，检验联通性，如图 6-22 所示。

```
C:\>ping 192.168.1.1

正在 Ping 192.168.1.1 具有 32 字节的数据：
来自 192.168.1.1 的回复: 字节=32 时间=18ms TTL=125
来自 192.168.1.1 的回复: 字节=32 时间=18ms TTL=125
来自 192.168.1.1 的回复: 字节=32 时间=18ms TTL=125
来自 192.168.1.1 的回复: 字节=32 时间=18ms TTL=125

192.168.1.1 的 Ping 统计信息:
    数据包: 已发送 = 4，已接收 = 4，丢失 = 0 (0% 丢失)，
往返行程的估计时间(以毫秒为单位):
    最短 = 18ms，最长 = 18ms，平均 = 18ms
```

图 6-22　从计算机 PC1 ping 计算机 PC2

通过查看路由表可以发现，把 OSPF 路由重分布到 RIP 后，RIP 路由域内路由器（边界路由器 R2 除外）的路由表中会出现路由来源为"R"、目的地址为 192.168.0.0 和 192.168.1.0 的路由，这是重分布

到 RIP 的 OSPF 的路由；把 RIP 路由重分布到 OSPF 后，OSPF 路由域内路由器（边界路由器 R2 除外）的路由表中会出现路由来源为"O E2"、目的地址为 10.0.0.0 和 10.0.1.0 的路由，这是重分布到 OSPF 的 RIP 的路由。通过检验联通性可以发现，计算机 PC1 和 PC2 实现了互通。

6.4.4 不同进程 OSPF 路由重分布的配置

本小节网络拓扑如图 6-23 所示，路由器 R1、R2 和 R3 的接口，以及计算机 PC1、PC2 和 PC3 的 IP 地址已经配置完成，要求配置不同进程 OSPF（为路由器 R1 的 f0/0、s0/1/0 接口和路由器 R2 的 f0/0、s0/1/0 接口配置 OSPF 进程 1，为路由器 R2 的 s0/1/1 接口和路由器 R3 的 f0/0、s0/1/0 接口配置 OSPF 进程 2）路由重分布，实现计算机 PC1、PC2 和 PC3 互通。

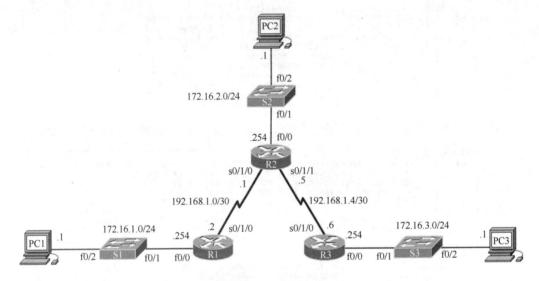

图 6-23 网络拓扑

步骤 1：在路由器 R1 的全局配置模式下配置 OSPF 进程 1，输入以下代码。

```
R1(config)#router ospf 1
R1(config-router)#network 192.168.1.0 0.0.0.3 area 0
R1(config-router)#network 172.16.1.0 0.0.0.255 area 0
```

步骤 2：在路由器 R2 的全局配置模式下配置 OSPF 进程 1 和进程 2，输入以下代码。

```
R2(config)#router ospf 1
R2(config-router)#network 172.16.2.0 0.0.0.255 area 0
R2(config-router)#network 192.168.1.0 0.0.0.3 area 0
R2(config-router)#router ospf 2
R2(config-router)#network 192.168.1.4 0.0.0.3 area 0
```

步骤 3：在路由器 R3 的全局配置模式下配置 OSPF 进程 2，输入以下代码。

```
R3(config)#router ospf 2
R3(config-router)#network 192.168.1.4 0.0.0.3 area 0
R3(config-router)#network 172.16.3.0 0.0.0.255 area 0
```

步骤 4：在路由器 R1 的特权执行模式下输入 show ip route，查看路由表，如图 6-24 所示。

```
R1#show ip route
Codes: C - connected, S - static, R - RIP, M - mobile, B - BGP
       D - EIGRP, EX - EIGRP external, O - OSPF, IA - OSPF inter area
       N1 - OSPF NSSA external type 1, N2 - OSPF NSSA external type 2
       E1 - OSPF external type 1, E2 - OSPF external type 2
       i - IS-IS, su - IS-IS summary, L1 - IS-IS level-1, L2 - IS-IS level-2
       ia - IS-IS inter area, * - candidate default, U - per-user static route
       o - ODR, P - periodic downloaded static route

Gateway of last resort is not set

     172.16.0.0/24 is subnetted, 2 subnets
C       172.16.1.0 is directly connected, FastEthernet0/0
O       172.16.2.0 [110/782] via 192.168.1.1, 00:00:18, Serial0/1/0
     192.168.1.0/24 is variably subnetted, 2 subnets, 2 masks
O       192.168.1.0/30 [110/1562] via 192.168.1.1, 00:00:18, Serial0/1/0
C       192.168.1.0/24 is directly connected, Serial0/1/0
```

图 6-24　路由器 R1 的路由表

步骤 5：在路由器 R2 的特权执行模式下输入 show ip route，查看路由表，如图 6-25 所示。

```
R2#show ip route
Codes: C - connected, S - static, R - RIP, M - mobile, B - BGP
       D - EIGRP, EX - EIGRP external, O - OSPF, IA - OSPF inter area
       N1 - OSPF NSSA external type 1, N2 - OSPF NSSA external type 2
       E1 - OSPF external type 1, E2 - OSPF external type 2
       i - IS-IS, su - IS-IS summary, L1 - IS-IS level-1, L2 - IS-IS level-2
       ia - IS-IS inter area, * - candidate default, U - per-user static route
       o - ODR, P - periodic downloaded static route

Gateway of last resort is not set

     172.16.0.0/24 is subnetted, 3 subnets
O       172.16.1.0 [110/782] via 192.168.1.2, 00:01:38, Serial0/1/0
C       172.16.2.0 is directly connected, FastEthernet0/0
O       172.16.3.0 [110/782] via 192.168.1.6, 00:02:15, Serial0/1/1
     192.168.1.0/24 is variably subnetted, 3 subnets, 2 masks
C       192.168.1.0/30 is directly connected, Serial0/1/0
O       192.168.1.0/24 [110/1562] via 192.168.1.2, 00:01:48, Serial0/1/0
C       192.168.1.4/30 is directly connected, Serial0/1/1
```

图 6-25　路由器 R2 的路由表

步骤 6：在路由器 R3 的特权执行模式下输入 show ip route，查看路由表，如图 6-26 所示。

```
R3#show ip route
Codes: C - connected, S - static, R - RIP, M - mobile, B - BGP
       D - EIGRP, EX - EIGRP external, O - OSPF, IA - OSPF inter area
       N1 - OSPF NSSA external type 1, N2 - OSPF NSSA external type 2
       E1 - OSPF external type 1, E2 - OSPF external type 2
       i - IS-IS, su - IS-IS summary, L1 - IS-IS level-1, L2 - IS-IS level-2
       ia - IS-IS inter area, * - candidate default, U - per-user static route
       o - ODR, P - periodic downloaded static route

Gateway of last resort is not set

     172.16.0.0/24 is subnetted, 1 subnets
C       172.16.3.0 is directly connected, FastEthernet0/0
     192.168.1.0/30 is subnetted, 1 subnets
C       192.168.1.4 is directly connected, Serial0/1/0
```

图 6-26　路由器 R3 的路由表

步骤 7：在计算机 PC1 的命令行界面输入 ping 172.16.2.1，检验联通性，如图 6-27 所示。

```
C:\>ping 172.16.2.1

正在 Ping 172.16.2.1 具有 32 字节的数据:
来自 172.16.2.1 的回复: 字节=32 时间=9ms TTL=126
来自 172.16.2.1 的回复: 字节=32 时间=9ms TTL=126
来自 172.16.2.1 的回复: 字节=32 时间=9ms TTL=126
来自 172.16.2.1 的回复: 字节=32 时间=9ms TTL=126

172.16.2.1 的 Ping 统计信息:
    数据包: 已发送 = 4, 已接收 = 4, 丢失 = 0 (0% 丢失),
往返行程的估计时间(以毫秒为单位):
    最短 = 9ms, 最长 = 9ms, 平均 = 9ms
```

图 6-27　从计算机 PC1 ping 计算机 PC2

步骤 8：在计算机 PC1 的命令行界面输入 ping 172.16.3.1，检验联通性，如图 6-28 所示。

```
C:\>ping 172.16.3.1

正在 Ping 172.16.3.1 具有 32 字节的数据：
来自 172.16.1.254 的回复：无法访问目标主机。
来自 172.16.1.254 的回复：无法访问目标主机。
来自 172.16.1.254 的回复：无法访问目标主机。
来自 172.16.1.254 的回复：无法访问目标主机。

172.16.3.1 的 Ping 统计信息：
    数据包：已发送 = 4，已接收 = 4，丢失 = 0 (0% 丢失)，
```

图 6-28　从计算机 PC1 ping 计算机 PC3

步骤 9：在计算机 PC2 的命令行界面输入 ping 172.16.3.1，检验联通性，如图 6-29 所示。

```
C:\>ping 172.16.3.1

正在 Ping 172.16.3.1 具有 32 字节的数据：
请求超时。
请求超时。
请求超时。
请求超时。

172.16.3.1 的 Ping 统计信息：
    数据包：已发送 = 4，已接收 = 0，丢失 = 4 (100% 丢失)，
```

图 6-29　从计算机 PC2 ping 计算机 PC3

步骤 10：把 OSPF 进程 2 的路由重分布到 OSPF 进程 1，在路由器 R2 的全局配置模式下输入以下代码。

```
R2(config)#router ospf 1
R2(config-router)#redistribute ospf 2 subnets
```

步骤 11：在路由器 R1 的特权执行模式下输入 show ip route，查看路由表，如图 6-30 所示。

```
R1#show ip route
Codes: C - connected, S - static, R - RIP, M - mobile, B - BGP
       D - EIGRP, EX - EIGRP external, O - OSPF, IA - OSPF inter area
       N1 - OSPF NSSA external type 1, N2 - OSPF NSSA external type 2
       E1 - OSPF external type 1, E2 - OSPF external type 2
       i - IS-IS, su - IS-IS summary, L1 - IS-IS level-1, L2 - IS-IS level-2
       ia - IS-IS inter area, * - candidate default, U - per-user static route
       o - ODR, P - periodic downloaded static route

Gateway of last resort is not set

     172.16.0.0/24 is subnetted, 3 subnets
C       172.16.1.0 is directly connected, FastEthernet0/0
O       172.16.2.0 [110/782] via 192.168.1.1, 00:03:47, Serial0/1/0
O E2    172.16.3.0 [110/782] via 192.168.1.1, 00:00:45, Serial0/1/0
     192.168.1.0/24 is variably subnetted, 3 subnets, 2 masks
O       192.168.1.0/30 [110/1562] via 192.168.1.1, 00:03:47, Serial0/1/0
C       192.168.1.0/24 is directly connected, Serial0/1/0
O E2    192.168.1.4/30 [110/781] via 192.168.1.1, 00:00:45, Serial0/1/0
```

图 6-30　路由器 R1 的路由表

步骤 12：把 OSPF 进程 1 的路由重分布到 OSPF 进程 2，在路由器 R2 的全局配置模式下输入以下代码。

```
R2(config-router)#router ospf 2
R2(config-router)#redistribute ospf 1 subnets
```

步骤 13：在路由器 R2 的特权执行模式下输入 show ip route，查看路由表，如图 6-31 所示。

```
R2#show ip route
Codes: C - connected, S - static, R - RIP, M - mobile, B - BGP
       D - EIGRP, EX - EIGRP external, O - OSPF, IA - OSPF inter area
       N1 - OSPF NSSA external type 1, N2 - OSPF NSSA external type 2
       E1 - OSPF external type 1, E2 - OSPF external type 2
       i - IS-IS, su - IS-IS summary, L1 - IS-IS level-1, L2 - IS-IS level-2
       ia - IS-IS inter area, * - candidate default, U - per-user static route
       o - ODR, P - periodic downloaded static route

Gateway of last resort is not set

     172.16.0.0/24 is subnetted, 3 subnets
O       172.16.1.0 [110/782] via 192.168.1.2, 00:04:13, Serial0/1/0
C       172.16.2.0 is directly connected, FastEthernet0/0
O       172.16.3.0 [110/782] via 192.168.1.6, 00:14:48, Serial0/1/1
     192.168.1.0/24 is variably subnetted, 3 subnets, 2 masks
C       192.168.1.0/30 is directly connected, Serial0/1/0
O       192.168.1.0/24 [110/1562] via 192.168.1.2, 00:14:20, Serial0/1/0
C       192.168.1.4/30 is directly connected, Serial0/1/1
```

图 6-31 路由器 R2 的路由表

步骤 14：在路由器 R3 的特权执行模式下输入 show ip route，查看路由表，如图 6-32 所示。

```
R3#show ip route
Codes: C - connected, S - static, R - RIP, M - mobile, B - BGP
       D - EIGRP, EX - EIGRP external, O - OSPF, IA - OSPF inter area
       N1 - OSPF NSSA external type 1, N2 - OSPF NSSA external type 2
       E1 - OSPF external type 1, E2 - OSPF external type 2
       i - IS-IS, su - IS-IS summary, L1 - IS-IS level-1, L2 - IS-IS level-2
       ia - IS-IS inter area, * - candidate default, U - per-user static route
       o - ODR, P - periodic downloaded static route

Gateway of last resort is not set

     172.16.0.0/24 is subnetted, 3 subnets
O E2    172.16.1.0 [110/782] via 192.168.1.5, 00:01:19, Serial0/1/0
O E2    172.16.2.0 [110/1] via 192.168.1.5, 00:01:19, Serial0/1/0
C       172.16.3.0 is directly connected, FastEthernet0/0
     192.168.1.0/24 is variably subnetted, 3 subnets, 2 masks
O E2    192.168.1.0/30 [110/781] via 192.168.1.5, 00:01:19, Serial0/1/0
O E2    192.168.1.0/24 [110/1562] via 192.168.1.5, 00:01:19, Serial0/1/0
C       192.168.1.4/30 is directly connected, Serial0/1/0
```

图 6-32 路由器 R3 的路由表

步骤 15：在计算机 PC1 的命令行界面输入 ping 172.16.2.1，检验联通性，如图 6-33 所示。

```
C:\>ping 172.16.2.1

正在 Ping 172.16.2.1 具有 32 字节的数据：
来自 172.16.2.1 的回复: 字节=32 时间=9ms TTL=126
来自 172.16.2.1 的回复: 字节=32 时间=9ms TTL=126
来自 172.16.2.1 的回复: 字节=32 时间=9ms TTL=126
来自 172.16.2.1 的回复: 字节=32 时间=9ms TTL=126

172.16.2.1 的 Ping 统计信息：
    数据包：已发送 = 4，已接收 = 4，丢失 = 0 (0% 丢失)，
往返行程的估计时间(以毫秒为单位)：
    最短 = 9ms，最长 = 9ms，平均 = 9ms
```

图 6-33 从计算机 PC1 ping 计算机 PC2

步骤 16：在计算机 PC1 的命令行界面输入 ping 172.16.3.1，检验联通性，如图 6-34 所示。

```
C:\>ping 172.16.3.1

正在 Ping 172.16.3.1 具有 32 字节的数据：
来自 172.16.3.1 的回复: 字节=32 时间=18ms TTL=125
来自 172.16.3.1 的回复: 字节=32 时间=18ms TTL=125
来自 172.16.3.1 的回复: 字节=32 时间=18ms TTL=125
来自 172.16.3.1 的回复: 字节=32 时间=18ms TTL=125

172.16.3.1 的 Ping 统计信息：
    数据包：已发送 = 4，已接收 = 4，丢失 = 0 (0% 丢失)，
往返行程的估计时间(以毫秒为单位)：
    最短 = 18ms，最长 = 18ms，平均 = 18ms
```

图 6-34 从计算机 PC1 ping 计算机 PC3

步骤 17：在计算机 PC2 的命令行界面输入 ping 172.16.3.1，检验联通性，如图 6-35 所示。

```
C:\>ping 172.16.3.1
正在 Ping 172.16.3.1 具有 32 字节的数据:
来自 172.16.3.1 的回复: 字节=32 时间=9ms TTL=126
来自 172.16.3.1 的回复: 字节=32 时间=9ms TTL=126
来自 172.16.3.1 的回复: 字节=32 时间=9ms TTL=126
来自 172.16.3.1 的回复: 字节=32 时间=9ms TTL=126

172.16.3.1 的 Ping 统计信息:
    数据包: 已发送 = 4,已接收 = 4,丢失 = 0 (0% 丢失),
往返行程的估计时间(以毫秒为单位):
    最短 = 9ms,最长 = 9ms,平均 = 9ms
```

图 6-35　从计算机 PC2 ping 计算机 PC3

通过查看路由表可以发现,当把 OSPF 进程 1 的路由和 OSPF 进程 2 的路由分别重分布后,路由器 R1 和 R3 的路由表中都出现了路由来源为"O E2"的路由,这是重分布到 OSPF 的 2 类外部路由。通过检验联通性可以发现,计算机 PC1、PC2 和 PC3 实现了互通。若为同一台路由器配置了不同的 OSPF 进程,运行两个不同 OSPF 进程的网络无法直接互通,就需要配置路由重分布,使两个进程相互学习彼此的路由信息,并使路由表中的路由完整,才能实现网络的互通。

6.5　项目小结

本项目完成了将静态路由重分布到 RIP 和 OSPF,RIP 和 OSPF 路由重分布,以及不同进程 OSPF 路由重分布的配置。当网络中配置了多种路由协议时,要实现网络互通,可以采用路由重分布;在同一台路由器上配置了不同的 OSPF 进程时,要实现网络互通,也需要配置路由重分布。

6.6　拓展训练

本项目的拓展训练网络拓扑如图 6-36 所示,某企业总公司网络运行的是 RIPv2,分公司网络运行的是 OSPF,总公司与分公司之间的网络采用静态路由,要求完成如下配置:

(1)完成路由器接口和计算机 IP 地址的配置;

(2)完成总公司 RIPv2、分公司 OSPF、静态路由和路由重分布的配置,实现计算机 PC1 和 PC2 互通。

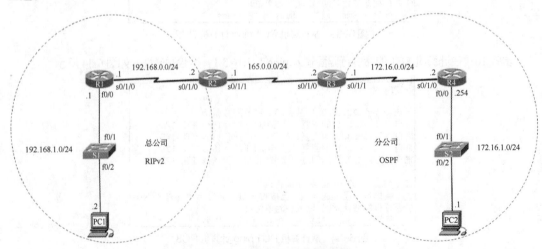

图 6-36　项目 6 拓展训练网络拓扑

模块二

网络交换的配置

交换机能够把同一局域网中的设备互联起来，分隔冲突域，并且通过 VLAN 对网络进行逻辑分组，抑制二层广播风暴。随着计算机网络技术的发展，出现了三层交换机，这种设备能够把二层交换功能和三层路由功能结合起来。在实际组网过程中，三层交换机的应用更加广泛。本模块主要介绍 VLAN 的配置、VLAN 中继的配置、VLAN 间路由的配置和三层交换机 VLAN 间路由的配置。

项目 7
VLAN 的配置

7.1 用户需求

某学校网络拓扑如图 7-1 所示。该学校网络由办公网络和学生网络组成，为了提高网络性能，要求办公网络和学生网络之间的广播相互隔离，怎样实现这个功能？

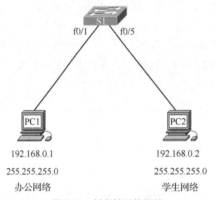

图 7-1 某学校网络拓扑

7.2 知识梳理

7.2.1 VLAN 的基本概念

虚拟局域网（Virtual Local Area Network，VLAN）是一组逻辑上联网的设备，VLAN 对局域网（Local Area Network，LAN）中的设备灵活地进行分段和组织，允许管理员根据功能、项目组或应用程序等因素划分网络，而不考虑用户或设备的物理位置。VLAN 创建逻辑广播域，每个 VLAN 都是一个独立的广播域，不同 VLAN 内的设备不能直接互通。VLAN 通过将大型广播域细分为较小网段来提高网络性能，并根据特定用户分组实施访问和安全策略。

VLAN 的基本概念和优点

7.2.2 VLAN 的优点

VLAN 主要具有安全性高、牺牲广播域换来性能提升，以及便于管理等优点。

1. 安全性高
含有敏感数据的用户组可与网络的其余部分隔离，从而降低泄露机密信息的可能性。
2. 牺牲广播域换来性能提升
将网络划分为多个 VLAN 可减少广播域中的设备数量，减少网络上不必要的流量，提升性能。
3. 便于管理
划分 VLAN 后，有相似网络需求的用户可以共享同一个 VLAN，让网络管理和应用管理更加轻松。

7.2.3 VLAN 的分类

VLAN 的分类和交换机的端口模式

1. 按照流量的类别分类

按流量的类别分类，VLAN 可分为数据 VLAN、默认 VLAN、本征 VLAN 和管理 VLAN 这 4 种。

（1）数据 VLAN

数据 VLAN 有时也称为用户 VLAN，用于传送用户生成的流量。数据 VLAN 用于将网络分为用户组和设备组。

（2）默认 VLAN

为交换机加载默认配置进行初始化后，所有交换机端口成为默认 VLAN 的一部分。加入默认 VLAN 的交换机端口属于同一个广播域。这样，连接到交换机任何端口的任何设备都能与连接到其他端口的其他设备通信。思科交换机的默认 VLAN 是 VLAN 1，所有端口都被分配给了 VLAN 1。默认情况下，所有的第 2 层控制流量都与 VLAN 1 关联，VLAN 1 具有所有 VLAN 的功能，它不能被重命名或删除。

（3）本征 VLAN

本征 VLAN 被分配给 IEEE 802.1Q trunk 端口，充当 trunk 链路两端的公共标识符，作用是维护无标记流量的向下兼容性。trunk 端口是交换机之间的端口，支持传输与多个 VLAN 关联的流量。IEEE 802.1Q trunk 端口支持来自多个 VLAN 的流量（有标记流量），也支持来自 VLAN 以外的流量（无标记流量）。在本征 VLAN（默认为 VLAN 1）中保存无标记流量。

（4）管理 VLAN

管理 VLAN 是用于访问交换机管理功能的 VLAN。创建交换机的管理 VLAN 时，需要为该交换机虚拟接口（SVI）分配 IP 地址和子网掩码，使交换机通过超文本传输协议（Hyper Text Transfer Protocol，HTTP）、Telnet、SSH 或简单网络管理协议（Simple Network Management Protocol，SNMP）进行管理。

2. 其他类别的 VLAN

常见的 VLAN 类别还有基于端口的 VLAN、基于协议的 VLAN、基于介质访问控制（Medium Access Control，MAC）地址的 VLAN 和基于 IP 子网的 VLAN。

（1）基于端口的 VLAN

基于端口的 VLAN 按照设备端口来定义成员。将指定端口加入指定 VLAN，该端口可以转发指定 VLAN 的数据帧。

（2）基于协议的 VLAN

基于协议的 VLAN 根据端口接收的帧所属的协议（族）类型及封装格式来给帧分配不同的 VLAN ID。可以用来划分 VLAN 的协议族有 IP、互联网络数据包交换（Internetwork Packet Exchange，IPX）和 AppleTalk 等。

（3）基于 MAC 地址的 VLAN

基于 MAC 地址的 VLAN 根据每个主机网卡的 MAC 地址来划分 VLAN，可以灵活移动主机，不受物理位置限制。

（4）基于 IP 子网的 VLAN

基于 IP 子网的 VLAN 以帧中 IP 报文的源 IP 地址及子网掩码作为依据来进行划分。设备从端口接收报文后，根据报文中的源 IP 地址找到与现有 VLAN 的对应关系，然后自动将其划分到指定 VLAN 中转发。

如果为交换机的某个端口同时配置了以上 4 种 VLAN，那么默认情况下，VLAN 将按照基于 MAC 地址的 VLAN、基于 IP 子网的 VLAN、基于协议的 VLAN、基于端口的 VLAN 的先后顺序进行匹配。

7.2.4 交换机的端口模式

交换机的端口模式主要有 access 和 trunk 这两种。

1. access 模式

access 模式的端口只能属于一个 VLAN，只允许特定的 VLAN 流量通过它发送的帧不带有 VLAN 标签，一般用于连接计算机的端口。

2. trunk 模式

trunk 模式的端口允许多个 VLAN 的流量通过，它发送的帧一般是带有 VLAN 标签的，一般用于交换机之间连接的端口。

7.2.5 配置命令

1. 创建 VLAN 并命名

```
Switch(config)#vlan vlan-id
Switch(config-vlan)#name vlan-name
```

（1）普通范围的 VLAN

vlan-id 的取值范围为 1～1005，用于中小型网络。其中，1002～1005 保留，供令牌环 VLAN 和 FDDI VLAN 使用。VLAN 的配置被存储在名为 vlan.dat 的 VLAN 数据库文件中，vlan.dat 文件位于交换机的闪存中。

（2）扩展范围的 VLAN

vlan-id 的取值范围为 1006～4094，可让服务提供商扩展自己的基础架构以满足更多客户的需求。某些跨国企业的规模很大，从而需要使用扩展范围的 VLAN，它支持的 VLAN 功能比普通范围的 VLAN 支持的更少，扩展范围的 VLAN 被保存在运行配置文件中，VLAN 中继协议（VLAN Trunking Protocol，VTP）无法识别扩展范围的 VLAN。

vlan-name：VLAN 的名称。

2. 为 VLAN 分配端口

（1）进入端口配置模式

```
Switch(config)#interface interface-id
```

（2）将端口设置为接入模式

```
Switch(config-if)#switchport mode access
```

（3）将端口分配到特定的 VLAN

```
Switch(config-if)#switchport access vlan vlan-id
```

3. 删除 VLAN

```
Switch(config)# no vlan vlan-id
```

注意 删除 VLAN 前,需要将所有属于该 VLAN 的端口移出。

4. 配置管理 VLAN

(1) 进入 SVI 配置模式

```
Switch(config)#interface vlan vlan-id
```

(2) 配置管理接口的 IP 地址

```
Switch(config-if)#ip address ip-address mask
```

ip-address:交换机的管理地址。

mask:子网掩码。

(3) 启用管理接口

```
Switch(config-if)#no shutdown
```

5. 查看 VLAN 命令

```
show vlan [brief|id vlan-id|name vlan-name|summary]
```

brief:每行显示一个 VLAN 的名称、状态和端口。

id *vlan-id*:显示由 VLAN ID 标识的某个 VLAN 的相关信息,*vlan-id* 的取值范围是 1~4094。

name *vlan-name*:显示由 VLAN 名称标识的某个 VLAN 的相关信息。

summary:显示 VLAN 摘要信息。

6. 查看端口命令

```
show interfaces [interface-id|vlan vlan-id]|switchport
```

interface-id:有效的端口,包括物理端口(包括类型、模块和端口号)和端口通道,端口通道的取值范围是 1~6。

vlan *vlan-id*:VLAN 标识符,取值范围是 1~4094。

switchport:显示交换端口的管理状态和运行状态,包括端口阻塞设置和端口保护设置。

7.3 方案设计

在图 7-1 所示的网络拓扑中,要实现办公网络和学生网络之间广播的相互隔离,可以在交换机上划分 VLAN,并把办公网络和学生网络互联的端口分配到不同的 VLAN 中。

7.4 项目实施

7.4.1 交换机 VLAN 的查看

本小节网络拓扑如图 7-1 所示,计算机 PC1 和 PC2 的 IP 地址已经配置完成,要求实现交换机 VLAN 的查看,并检验计算机 PC1 和 PC2 的联通性。

步骤 1:在交换机 S1 的特权执行模式下输入 show vlan,查看交换机 S1 的 VLAN 信息,如图 7-2 所示。

步骤 2:在计算机 PC1 的命令行界面输入 ping 192.168.0.2,检验联通性,如图 7-3 所示。

```
S1#show vlan

VLAN Name                             Status     Ports
---- -------------------------------- ---------  -------------------------------
1    default                          active     Fa0/1, Fa0/2, Fa0/3, Fa0/4
                                                 Fa0/5, Fa0/6, Fa0/7, Fa0/8
                                                 Fa0/9, Fa0/10, Fa0/11, Fa0/12
                                                 Fa0/13, Fa0/14, Fa0/15, Fa0/16
                                                 Fa0/17, Fa0/18, Fa0/19, Fa0/20
                                                 Fa0/21, Fa0/22, Fa0/23, Fa0/24
                                                 Gi0/1, Gi0/2
1002 fddi-default                     act/unsup
1003 token-ring-default               act/unsup
1004 fddinet-default                  act/unsup
1005 trnet-default                    act/unsup

VLAN Type  SAID     MTU   Parent RingNo BridgeNo Stp  BrdgMode Trans1 Trans2
---- ----- -------- ----- ------ ------ -------- ---- -------- ------ ------
1    enet  100001   1500  -      -      -        -             0      0
1002 fddi  101002   1500  -      -      -        -             0      0
1003 tr    101003   1500  -      -      -        -             0      0
1004 fdnet 101004   1500  -      -      -        ieee -        0      0
1005 trnet 101005   1500  -      -      -        ibm  -        0      0
```

图 7-2　交换机 S1 的 VLAN 信息

```
C:\>ping 192.168.0.2

正在 Ping 192.168.0.2 具有 32 字节的数据:
来自 192.168.0.2 的回复: 字节=32 时间<1ms TTL=128
来自 192.168.0.2 的回复: 字节=32 时间<1ms TTL=128
来自 192.168.0.2 的回复: 字节=32 时间<1ms TTL=128
来自 192.168.0.2 的回复: 字节=32 时间<1ms TTL=128

192.168.0.2 的 Ping 统计信息:
    数据包: 已发送 = 4，已接收 = 4，丢失 = 0 (0% 丢失),
往返行程的估计时间(以毫秒为单位):
    最短 = 0ms, 最长 = 0ms, 平均 = 0ms
```

图 7-3　从计算机 PC1 ping 计算机 PC2

通过查看交换机 S1 的 VLAN 信息和检验联通性可以发现，在默认情况下，交换机的所有端口都在 VLAN 1 内，接入交换机的计算机可以直接互通。

7.4.2　交换机 VLAN 的配置

本小节网络拓扑如图 7-1 所示，要求完成交换机 VLAN 的配置（将交换机 S1 的 f0/1～f0/4 端口划分到 VLAN 10，命名为 office，将交换机 S1 的 f0/5～f0/8 端口划分到 VLAN 20，命名为 student），实现办公网络和学生网络之间广播的相互隔离。

步骤 1：创建 VLAN，在交换机 S1 的全局配置模式下输入以下代码。

```
S1(config)#vlan 10
S1(config-vlan)#name office
S1(config-vlan)#vlan 20
S1(config-vlan)#name student
```

步骤 2：为 VLAN 分配端口，在交换机 S1 的全局配置模式下输入以下代码。

```
S1(config)#interface f0/1
S1(config-if)#switchport mode access
S1(config-if)#switchport access vlan 10
S1(config-if)#interface range f0/2-4
S1(config-if-range)#switchport mode access
```

```
S1(config-if-range)#switchport access vlan 10
S1(config-if-range)#interface range f0/5-8
S1(config-if-range)#switchport mode access
S1(config-if-range)#switchport access vlan 20
```

当需要把多个端口分配给同一个 VLAN 时，可以在 interface 后面加 range 关键字，并在后面输入端口号的范围，这样可以把多个端口同时分配给某个 VLAN。

步骤 3：在交换机 S1 的特权执行模式下输入 show vlan，查看交换机 S1 的 VLAN 信息，如图 7-4 所示。

```
S1#show vlan

VLAN Name                          Status    Ports
---- ------------------------------ --------- -------------------------------
1    default                        active    Fa0/9, Fa0/10, Fa0/11, Fa0/12
                                              Fa0/13, Fa0/14, Fa0/15, Fa0/16
                                              Fa0/17, Fa0/18, Fa0/19, Fa0/20
                                              Fa0/21, Fa0/22, Fa0/23, Fa0/24
                                              Gi0/1, Gi0/2
10   VLAN0010                       active    Fa0/1, Fa0/2, Fa0/3, Fa0/4
20   VLAN0020                       active    Fa0/5, Fa0/6, Fa0/7, Fa0/8
1002 fddi-default                   act/unsup
1003 token-ring-default             act/unsup
1004 fddinet-default                act/unsup
1005 trnet-default                  act/unsup

VLAN Type  SAID       MTU   Parent RingNo BridgeNo Stp  BrdgMode Trans1 Trans2
---- ----- ---------- ----- ------ ------ -------- ---- -------- ------ ------
1    enet  100001     1500  -      -      -        -    -        0      0
10   enet  100010     1500  -      -      -        -    -        0      0
20   enet  100020     1500  -      -      -        -    -        0      0
1002 fddi  101002     1500  -      -      -        -    -        0      0
1003 tr    101003     1500  -      -      -        -    -        0      0
1004 fdnet 101004     1500  -      -      -        ieee -        0      0
1005 trnet 101005     1500  -      -      -        ibm  -        0      0
```

图 7-4　交换机 S1 的 VLAN 信息

步骤 4：在计算机 PC1 的命令行界面输入 ping 192.168.0.2，检验联通性，如图 7-5 所示。

```
C:\>ping 192.168.0.2

正在 Ping 192.168.0.2 具有 32 字节的数据:
来自 192.168.0.1 的回复: 无法访问目标主机。
来自 192.168.0.1 的回复: 无法访问目标主机。
来自 192.168.0.1 的回复: 无法访问目标主机。
来自 192.168.0.1 的回复: 无法访问目标主机。

192.168.0.2 的 Ping 统计信息:
    数据包: 已发送 = 4，已接收 = 4，丢失 = 0 (0% 丢失)，
```

图 7-5　从计算机 PC1 ping 计算机 PC2

通过查看交换机 S1 的 VLAN 信息和检验联通性可以发现，当为交换机划分了 VLAN 后，即使是连接同一个交换机的计算机，只要计算机所连接的交换机端口在不同的 VLAN 中，计算机也不会直接互通。

7.5　项目小结

本项目完成了交换机 VLAN 的配置，在默认配置下，思科交换机的所有端口都被分配给了 VLAN 1，接入交换机端口的计算机可以直接互通。当为交换机配置 VLAN 后，接入计算机的端口如果位于不同的 VLAN，则计算机不能直接互通。

7.6 拓展训练

本项目的拓展训练网络拓扑如图 7-6 所示，要求完成如下配置：
（1）在交换机 S1 上创建 VLAN 10；
（2）将交换机的 f0/1 端口分配给 VLAN 10；
（3）为交换机 S1 配置管理地址，管理 VLAN 为 VLAN 10，使计算机 PC1 能够 ping 通交换机 S1。

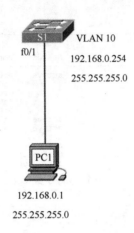

图 7-6　项目 7 拓展训练网络拓扑

项目 8
VLAN中继的配置

8.1 用户需求

某学校综合楼网络拓扑如图 8-1 所示。该综合楼网络由办公网络和学生网络组成，交换机 S1 位于 2 楼配线间，交换机 S2 位于 4 楼配线间。怎样实现交换机 S1 连接的办公网络与交换机 S2 连接的办公网络，以及交换机 S1 连接的学生网络与交换机 S2 连接的学生网络互通？

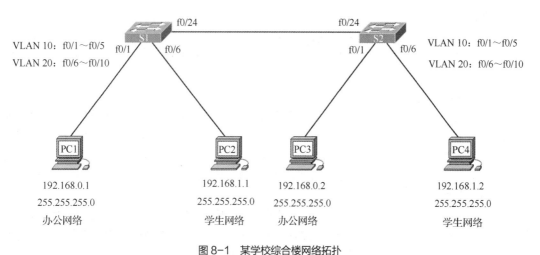

图 8-1 某学校综合楼网络拓扑

8.2 知识梳理

8.2.1 VLAN 中继

VLAN 中继建立在两台网络设备之间的点对点链路上，负责传输多个 VLAN 的流量。VLAN 中继可以让 VLAN 扩展到整个网络，思科设备支持 IEEE 802.1Q。

VLAN 中继

8.2.2 帧在中继链路上的转发

当通过中继链路转发帧前，交换机会将 VLAN ID 标记添加到帧；当通过非中继链路转发帧前，交换机会删除其中的标记。添加了 VLAN ID 标记的帧可以通过中继链路经过任意数量的交换机,在目的地正确的 VLAN 中进行转发。当中继端口收到无标记帧时,交换机会将该帧转发给本征 VLAN。

> **边学边思**
>
> 有了 VLAN 中继,位于不同交换机、相同 VLAN 的计算机就可以成功通信,可见 VLAN 中继在该功能的实现中发挥了"桥梁"的作用。有效的团队合作可以将不同人的技能和专长结合起来,共同解决问题和完成任务。分工合作和互相支持,可以提高团队工作效率,节约时间和其他资源,从而实现更好的效果。

8.2.3 中继模式

中继模式定义了端口与其对等端口如何使用动态中继协议(Dynamic Trunking Protocol,DTP)来协商并建立中继链路。DTP 是思科专有的协议。当为交换机端口配置了某些中继模式后,此端口会自动启用 DTP。DTP 可以管理中继协商,但前提是为另一台交换机的端口配置了某个支持 DTP 的中继模式。DTP 同时支持 ISL(Interior Switching Link)中继和 IEEE 802.1Q 中继,思科交换机的端口支持的中继模式有以下 3 种。

1. 开启模式

开启模式(On)下,本地的交换机端口定期向远程端口发送 DTP 帧(通告),本地的交换机端口通告远程端口它正在动态地进入中继状态。不管远程端口发出何种 DTP 帧作为对通告的响应,本地端口都会进入中继状态。思科交换机端口默认的中继模式为开启模式。

2. 自动模式

自动模式(Auto)下,本地的交换机端口定期向远程端口发送 DTP 帧,本地的交换机端口通告远程交换机端口它能够进入中继状态,但是没有请求进入中继状态。经过 DTP 协商后,仅当远程端口中继模式已被配置为开启或期望模式时,本地端口才最终进入中继状态。如果本地的交换机端口和远程端口都被设置为自动模式,则它们不会协商进入中继状态,而是协商进入接入状态(非中继状态)。

3. 期望模式

期望模式(Desirable)下,本地的交换机端口定期向远程端口发送 DTP 帧,本地的交换机端口通告远程交换机端口它能够进入中继状态,并请求远程交换机端口进入中继状态。如果本地端口检测到远程端口已被配置为开启、期望或自动模式,则本地端口最终进入中继状态。

8.2.4 VLAN 中继的配置

1. 通过特定交换机端口进入接口配置模式

```
Switch(config)#interface interface-id
```

2. 配置交换机的 trunk 模式

```
Switch(config-if)#switchport trunk encapsulation [dot1q|isl]
```

如果交换机支持多种 trunk 模式,则可以通过该命令配置 trunk 模式。

3. 强制连接交换机端口的链路成为中继(trunk)链路

```
Switch(config-if)#switchport mode trunk
```

4. 配置本征 VLAN

```
Switch(config-if)#switchport trunk native vlan vlan-id
```

将 VLAN 指定为本征 VLAN，用于为 IEEE 802.1Q 中继传输无标记流量。

5. 配置中继链路上允许通过的 VLAN

```
Switch(config-if)#switchport trunk allowed vlan {all|[add|remove|except]} vlan-list
```

all：许可 VLAN 列表，包含所有支持的 VLAN。

add：将指定 VLAN 列表加入许可 VLAN 列表。

remove：将指定 VLAN 列表从许可 VLAN 列表中删除。

except：将除列出的 VLAN 外的其他 VLAN 加入许可 VLAN 列表。

6. 将中继链路上允许通过的 VLAN 和本征 VLAN 恢复为默认状态

```
Switch(config-if)#no switchport trunk allowed vlan
Switch(config-if)#no switchport trunk native vlan
```

7. 配置中继模式为自动模式

```
Switch(config)#interface interface-id
Switch(config-if)#switchport mode dynamic auto
```

8. 配置中继模式为期望模式

```
Switch(config)#interface interface-id
Switch(config-if)#switchport mode dynamic desirable
```

9. 查看交换机端口的配置

```
Switch#show interfaces interface-id switchport
```

该命令可以用于检验中继配置。

8.3 方案设计

在图 8-1 所示的网络拓扑中，如果将两台交换机互联的端口配置为 access 模式，那么这条链路只能传输一个 VLAN 的流量。要实现交换机 S1 互联的办公网络与交换机 S2 互联的办公网络，以及交换机 S1 互联的学生网络与交换机 S2 互联的学生网络互通，可以把交换机 S1 和 S2 互联的链路配置成中继链路。

8.4 项目实施

根据图 8-1 所示的网络拓扑，计算机的 IP 地址已经配置完成，交换机 S1 和 S2 的 VLAN 按照网络拓扑中的标注配置。要求完成 VLAN 和 VLAN 中继的配置，实现交换机 S1 互联的办公网络与交换机 S2 互联的办公网络，以及交换机 S1 互联的学生网络与交换机 S2 互联的学生网络互通。

步骤 1：在交换机 S1 的全局配置模式下创建 VLAN，输入以下代码。

```
S1(config)#vlan 10
S1(config-vlan)#vlan 20
```

步骤 2：在交换机 S2 的全局配置模式下创建 VLAN，输入以下代码。

```
S2(config)#vlan 10
S2(config-vlan)#vlan 20
```

步骤 3：在交换机 S1 的全局配置模式下为 VLAN 分配端口，输入以下代码。

```
S1(config-if)#interface rang f0/1-5
S1(config-if-range)#switchport mode access
```

```
S1(config-if-range)#switchport access vlan 10
S1(config-if-range)#interface range f0/6-10
S1(config-if-range)#switchport mode access
S1(config-if-range)#switchport access vlan 20
```

步骤 4：在 S2 的全局配置模式下为 VLAN 分配端口，输入以下代码。

```
S2(config-if)#interface rang f0/1-5
S2(config-if-range)#switchport mode access
S2(config-if-range)#switchport access vlan 10
S2(config-if-range)#interface range f0/6-10
S2(config-if-range)#switchport mode access
S2(config-if-range)#switchport access vlan 20
```

步骤 5：在交换机 S1 的特权执行模式下输入 show vlan，查看交换机 S1 的 VLAN 信息，如图 8-2 所示。

```
S1#show vlan

VLAN Name                             Status    Ports
---- -------------------------------- --------- -------------------------------
1    default                          active    Fa0/11, Fa0/12, Fa0/13, Fa0/14
                                                Fa0/15, Fa0/16, Fa0/17, Fa0/18
                                                Fa0/19, Fa0/20, Fa0/21, Fa0/22
                                                Fa0/23, Fa0/24, Gi0/1, Gi0/2
10   VLAN0010                         active    Fa0/1, Fa0/2, Fa0/3, Fa0/4
                                                Fa0/5
20   VLAN0020                         active    Fa0/6, Fa0/7, Fa0/8, Fa0/9
                                                Fa0/10
1002 fddi-default                     act/unsup
1003 token-ring-default               act/unsup
1004 fddinet-default                  act/unsup
1005 trnet-default                    act/unsup

VLAN Type  SAID       MTU   Parent RingNo BridgeNo Stp  BrdgMode Trans1 Trans2
---- ----- ---------- ----- ------ ------ -------- ---- -------- ------ ------
1    enet  100001     1500  -      -      -        -    -        0      0
10   enet  100010     1500  -      -      -        -    -        0      0
20   enet  100020     1500  -      -      -        -    -        0      0
1002 fddi  101002     1500  -      -      -        -    -        0      0
1003 tr    101003     1500  -      -      -        -    -        0      0
1004 fdnet 101004     1500  -      -      -        ieee -        0      0
1005 trnet 101005     1500  -      -      -        ibm  -        0      0
```

图 8-2　交换机 S1 的 VLAN 信息

步骤 6：在交换机 S2 的特权执行模式下输入 show vlan，查看交换机 S2 的 VLAN 信息，如图 8-3 所示。

```
S2#show vlan

VLAN Name                             Status    Ports
---- -------------------------------- --------- -------------------------------
1    default                          active    Fa0/11, Fa0/12, Fa0/13, Fa0/14
                                                Fa0/15, Fa0/16, Fa0/17, Fa0/18
                                                Fa0/19, Fa0/20, Fa0/21, Fa0/22
                                                Fa0/23, Fa0/24, Gi0/1, Gi0/2
10   VLAN0010                         active    Fa0/1, Fa0/2, Fa0/3, Fa0/4
                                                Fa0/5
20   VLAN0020                         active    Fa0/6, Fa0/7, Fa0/8, Fa0/9
                                                Fa0/10
1002 fddi-default                     act/unsup
1003 token-ring-default               act/unsup
1004 fddinet-default                  act/unsup
1005 trnet-default                    act/unsup

VLAN Type  SAID       MTU   Parent RingNo BridgeNo Stp  BrdgMode Trans1 Trans2
---- ----- ---------- ----- ------ ------ -------- ---- -------- ------ ------
1    enet  100001     1500  -      -      -        -    -        0      0
10   enet  100010     1500  -      -      -        -    -        0      0
20   enet  100020     1500  -      -      -        -    -        0      0
1002 fddi  101002     1500  -      -      -        -    -        0      0
1003 tr    101003     1500  -      -      -        -    -        0      0
1004 fdnet 101004     1500  -      -      -        ieee -        0      0
1005 trnet 101005     1500  -      -      -        ibm  -        0      0
```

图 8-3　交换机 S2 的 VLAN 信息

步骤 7：在交换机 S1 的全局配置模式下将 f0/24 端口配置为 trunk 模式，输入以下代码。

```
S1(config)#interface f0/24
```

```
S1(config-if)#sw mode trunk
S1(config-if)#sw trunk allowed vlan 10,20
```

步骤 8：在交换机 S2 的全局配置模式下将 f0/24 端口配置为 trunk 模式，输入以下代码。

```
S2(config)#interface f0/24
S2(config-if)#sw mode trunk
S2(config-if)#sw trunk allowed vlan 10,20
```

步骤 9：在计算机 PC1 的命令行界面输入 ping 192.168.0.2，检验联通性，如图 8-4 所示。

```
C:\>ping 192.168.0.2

正在 Ping 192.168.0.2 具有 32 字节的数据：
来自 192.168.0.2 的回复: 字节=32 时间<1ms TTL=128
来自 192.168.0.2 的回复: 字节=32 时间<1ms TTL=128
来自 192.168.0.2 的回复: 字节=32 时间<1ms TTL=128
来自 192.168.0.2 的回复: 字节=32 时间<1ms TTL=128

192.168.0.2 的 Ping 统计信息：
    数据包: 已发送 = 4，已接收 = 4，丢失 = 0 (0% 丢失)，
往返行程的估计时间(以毫秒为单位)：
    最短 = 0ms，最长 = 0ms，平均 = 0ms
```

图 8-4　从计算机 PC1 ping 计算机 PC3

步骤 10：在计算机 PC1 的命令行界面输入 ping 192.168.1.2，检验联通性，如图 8-5 所示。

```
C:\>ping 192.168.1.2

正在 Ping 192.168.1.2 具有 32 字节的数据：
PING: 传输失败。常见故障。
PING: 传输失败。常见故障。
PING: 传输失败。常见故障。
PING: 传输失败。常见故障。

192.168.1.2 的 Ping 统计信息：
    数据包: 已发送 = 4，已接收 = 0，丢失 = 4 (100% 丢失)，
```

图 8-5　从计算机 PC1 ping 计算机 PC4

步骤 11：在计算机 PC2 的命令行界面输入 ping 192.168.0.2，检验联通性，如图 8-6 所示。

```
C:\>ping 192.168.0.2

正在 Ping 192.168.0.2 具有 32 字节的数据：
PING: 传输失败。常见故障。
PING: 传输失败。常见故障。
PING: 传输失败。常见故障。
PING: 传输失败。常见故障。

192.168.0.2 的 Ping 统计信息：
    数据包: 已发送 = 4，已接收 = 0，丢失 = 4 (100% 丢失)，
```

图 8-6　从计算机 PC2 ping 计算机 PC3

步骤 12：在计算机 PC2 的命令行界面输入 ping 192.168.1.2，检验联通性，如图 8-7 所示。

```
C:\>ping 192.168.1.2
正在 Ping 192.168.1.2 具有 32 字节的数据:
来自 192.168.1.2 的回复: 字节=32 时间=1ms TTL=128
来自 192.168.1.2 的回复: 字节=32 时间<1ms TTL=128
来自 192.168.1.2 的回复: 字节=32 时间=1ms TTL=128
来自 192.168.1.2 的回复: 字节=32 时间<1ms TTL=128

192.168.1.2 的 Ping 统计信息:
    数据包: 已发送 = 4,已接收 = 4,丢失 = 0 (0% 丢失),
往返行程的估计时间(以毫秒为单位):
    最短 = 0ms,最长 = 1ms,平均 = 0ms
```

图 8-7　从计算机 PC2 ping 计算机 PC4

通过查看交换机的 VLAN 信息和检验联通性可以发现,中继链路上可以传输多个 VLAN 的流量,可以实现位于不同交换机、相同 VLAN 的计算机互通。中继链路不能实现位于同一交换机、不同 VLAN 的计算机互通。

8.5　项目小结

本项目完成了 VLAN 中继的配置。要实现位于不同交换机、相同 VLAN 的计算机互通,可以把交换机互联的链路配置成中继链路,这样一条链路可以传输多个 VLAN 的流量,也可以配置中继链路上允许通过的 VLAN。

8.6　拓展训练

本项目的拓展训练网络拓扑如图 8-8 所示,要求完成如下配置:
(1)根据图 8-8 中的标注,为交换机 S1 和 S3 创建 VLAN,并把端口分配给 VLAN;
(2)完成 VLAN 中继的配置和交换机 S2 必要的 VLAN 配置,实现计算机 PC1 和 PC3 互通、计算机 PC2 和 PC4 互通。

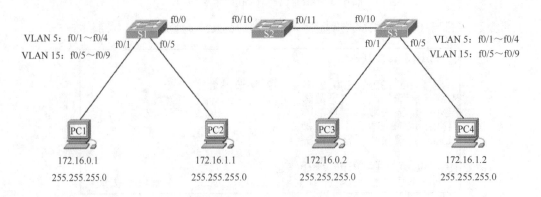

图 8-8　项目 8 拓展训练网络拓扑

项目 9
VLAN间路由的配置

9.1 用户需求

某学校网络拓扑如图 9-1 所示。两台计算机位于不同的 VLAN，怎样才能实现位于不同 VLAN 的计算机互通？

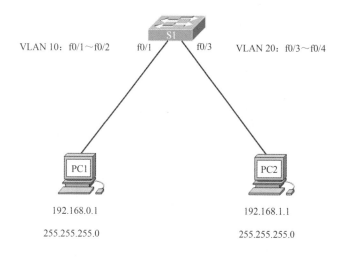

图 9-1 某学校网络拓扑

9.2 知识梳理

9.2.1 VLAN 间路由

VLAN 用于分段交换网络、分隔广播域，没有路由器的参与，通过位于不同 VLAN 内的交换机端口互联的计算机无法通信。任何支持三层路由功能的设备（如路由器或多层交换机）都可以将网络流量从一个 VLAN 转发至另一个 VLAN，这一转发过程称为 VLAN 间路由。

VLAN 间路由

9.2.2 传统方式的 VLAN 间路由

传统方式的 VLAN 间路由是把路由器的物理接口连接到交换机的物理端口，被分配到不同 VLAN 的交换机端口以接入模式连接到路由器，路由器的每个接口都配置一个 IP 地址，该 IP 地址与所连接的特定 VLAN 子网相关联，与 VLAN 相连的网络设备可通过连接到同一 VLAN 的物理接口与路由器通信。路由器通过物理接口连接到唯一的 VLAN，从而实现流量转发。传统方式的 VLAN 间路由要求路由器和交换机都必须有多个物理接口。

9.2.3 单臂路由

单臂路由是指在路由器的单个物理接口上通过配置子接口的方式，实现原来相互隔离的不同 VLAN 之间的互联互通。将交换机互联路由器的端口配置成 trunk 模式，通过接收中继链路上来自相邻交换机的 VLAN 的标记流量，实现子接口在 VLAN 之间的内部路由，进而实现 VLAN 间路由。

子接口是与同一物理接口相关联的虚拟接口。在路由器的软件中配置子接口，使子接口能够在特定的 VLAN 上运行。根据各自的 VLAN 分配，子接口被配置到不同的子网，以便被标记了 VLAN 的流量在物理接口发回之前先进行逻辑路由。

9.2.4 配置命令

1. 开启物理接口

```
Router(config)#interface interface-id
Router(config-if)#no shutdown
```

2. 创建子接口

```
Router(config)#interface interface-id.subinterface-id
```

interface-id.subinterface-id：物理接口.子接口的编号，子接口的语法为物理接口加上一个点，再加上子接口的编号。通常情况下，子接口的编号与该子接口传输的 VLAN ID 一致。

3. 封装 dot1Q

```
Router(config-subif)#encapsulation dot1Q vlan-id
```

vlan-id：指定子接口传递的 VLAN ID，这个参数必须与该子接口要传递的 VLAN ID 一致。

4. 配置子接口的 IP 地址

```
Router(config-subif)#ip address ip-address mask
```

ip-address：子接口的 IP 地址。
mask：子接口的 IP 地址对应的子网掩码。

9.3 方案设计

在图 9-1 所示的网络拓扑中，位于 VLAN 中的设备在逻辑功能上与位于物理上的 VLAN 中的设备相同，即 VLAN 中的设备的逻辑功能与它们的物理位置无关。要实现位于不同 VLAN 的计算机互通，需要使用三层设备，可以采用传统方式的 VLAN 间路由来实现，也可以采用单臂路由来实现。

采用单臂路由时，路由器的一个物理接口连接的链路要传递多个 VLAN 的流量，需将与路由器互连的交换机的端口配置成 trunk 模式。

9.4 项目实施

9.4.1 传统方式 VLAN 间路由的配置

本小节网络拓扑如图 9-2 所示。计算机的 IP 地址已经配置完成，要求完成传统方式 VLAN 间路由的配置，实现计算机 PC1 和 PC2 互通。

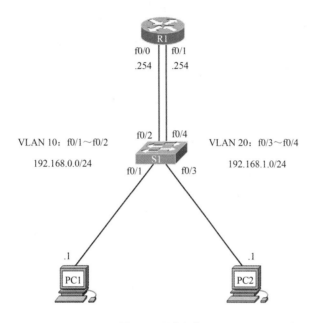

图 9-2　网络拓扑

步骤 1：在交换机 S1 的全局配置模式下配置 VLAN，输入以下代码。

```
S1(config)#vlan 10
S1(config-vlan)#vlan 20
S1(config-vlan)#interface range f0/1-2
S1(config-if-range)#sw mode acc
S1(config-if-range)#sw acc vlan 10
S1(config-if-range)#interface range f0/3-4
S1(config-if-range)#sw mode acc
S1(config-if-range)#sw acc vlan 20
```

步骤 2：在路由器 R1 的全局配置模式下配置接口，输入以下代码。

```
R1(config)#interface f0/0
R1(config-if)#ip add 192.168.0.254 255.255.255.0
R1(config-if)#no shutdown
R1(config-if)#interface f0/1
R1(config-if)#ip add 192.168.1.254 255.255.255.0
R1(config-if)#no shutdown
```

步骤 3：在交换机 S1 的特权执行模式下输入 show vlan，查看交换机 S1 的 VLAN 信息，如图 9-3 所示。

```
S1#show vlan
VLAN Name                             Status    Ports
---- -------------------------------- --------- -------------------------------
1    default                          active    Fa0/5, Fa0/6, Fa0/7, Fa0/8
                                                Fa0/9, Fa0/10, Fa0/11, Fa0/12
                                                Fa0/13, Fa0/14, Fa0/15, Fa0/16
                                                Fa0/17, Fa0/18, Fa0/19, Fa0/20
                                                Fa0/21, Fa0/22, Fa0/23, Fa0/24
                                                Gi0/1, Gi0/2
10   VLAN0010                         active    Fa0/1, Fa0/2
20   VLAN0020                         active    Fa0/3, Fa0/4
1002 fddi-default                     act/unsup
1003 token-ring-default               act/unsup
1004 fddinet-default                  act/unsup
1005 trnet-default                    act/unsup

VLAN Type  SAID       MTU   Parent RingNo BridgeNo Stp  BrdgMode Trans1 Trans2
---- ----- ---------- ----- ------ ------ -------- ---- -------- ------ ------
1    enet  100001     1500  -      -      -        -    -        0      0
10   enet  100010     1500  -      -      -        -    -        0      0
20   enet  100020     1500  -      -      -        -    -        0      0
1002 fddi  101002     1500  -      -      -        -    -        0      0
1003 tr    101003     1500  -      -      -        -    -        0      0
1004 fdnet 101004     1500  -      -      -        ieee -        0      0
1005 trnet 101005     1500  -      -      -        ibm  -        0      0
```

图 9-3　交换机 S1 的 VLAN 信息

步骤 4：在路由器 R1 的特权执行模式下输入 show ip route，查看路由表，如图 9-4 所示。

```
R1#show ip route
Codes: C - connected, S - static, R - RIP, M - mobile, B - BGP
       D - EIGRP, EX - EIGRP external, O - OSPF, IA - OSPF inter area
       N1 - OSPF NSSA external type 1, N2 - OSPF NSSA external type 2
       E1 - OSPF external type 1, E2 - OSPF external type 2
       i - IS-IS, su - IS-IS summary, L1 - IS-IS level-1, L2 - IS-IS level-2
       ia - IS-IS inter area, * - candidate default, U - per-user static route
       o - ODR, P - periodic downloaded static route

Gateway of last resort is not set

C    192.168.0.0/24 is directly connected, FastEthernet0/0
C    192.168.1.0/24 is directly connected, FastEthernet0/1
```

图 9-4　路由器 R1 的路由表

步骤 5：在计算机 PC1 的命令行界面输入 ping 192.168.1.1，检验联通性，如图 9-5 所示。

```
C:\>ping 192.168.1.1

正在 Ping 192.168.1.1 具有 32 字节的数据：
来自 192.168.1.1 的回复：字节=32 时间<1ms TTL=127
来自 192.168.1.1 的回复：字节=32 时间<1ms TTL=127
来自 192.168.1.1 的回复：字节=32 时间<1ms TTL=127
来自 192.168.1.1 的回复：字节=32 时间<1ms TTL=127

192.168.1.1 的 Ping 统计信息：
    数据包：已发送 = 4，已接收 = 4，丢失 = 0 (0% 丢失)，
往返行程的估计时间(以毫秒为单位)：
    最短 = 0ms，最长 = 0ms，平均 = 0ms
```

图 9-5　从计算机 PC1 ping 计算机 PC2

通过查看网络拓扑和 VLAN 信息可以发现，本小节中路由器的 f0/0 接口连接到 VLAN 10，路由器的 f0/1 接口连接到 VLAN 20。通过查看路由表可以发现，当路由器的接口配置完成后，路由表中出现了直连路由。通过检验网络的联通性可以发现，计算机 PC1 和 PC2 实现了互通。

9.4.2 单臂路由的配置

本小节网络拓扑如图 9-6 所示，计算机的 IP 地址已经配置完成，要求完成单臂路由的配置，实现计算机 PC1 和 PC2 互通。

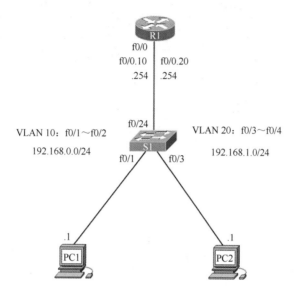

图 9-6 网络拓扑

步骤 1：在交换机 S1 的全局配置模式下配置 VLAN，输入以下代码。

```
S1(config)#vlan 10
S1(config-vlan)#vlan 20
S1(config-vlan)#interface range f0/1-2
S1(config-if-range)#sw mode acc
S1(config-if-range)#sw acc vlan 10
S1(config-if-range)#interface range f0/3-4
S1(config-if-range)#sw mode acc
S1(config-if-range)#sw acc vlan 20
```

步骤 2：在路由器 R1 的全局配置模式下配置子接口，输入以下代码。

```
R1(config)#interface f0/0
R1(config-if)#no shutdown
R1(config-if)#interface f0/0.10
R1(config-subif)#encapsulation dot1Q 10
R1(config-subif)#ip add 192.168.0.254 255.255.255.0
R1(config-subif)#interface f0/0.20
R1(config-subif)#encapsulation dot1Q 20
R1(config-subif)#ip add 192.168.1.254 255.255.255.0
```

步骤 3：在交换机 S1 的全局配置模式下把 f0/24 端口配置为 trunk 模式，输入以下代码。

```
S1(config)#interface f0/24
```

```
S1(config-if)#sw mode trunk
S1(config-if)#sw trunk allowed vlan 10,20
```

步骤 4：在路由器 R1 的特权执行模式下输入 show ip route，查看路由表，如图 9-7 所示。

```
R1#show ip route
Codes: C - connected, S - static, R - RIP, M - mobile, B - BGP
       D - EIGRP, EX - EIGRP external, O - OSPF, IA - OSPF inter area
       N1 - OSPF NSSA external type 1, N2 - OSPF NSSA external type 2
       E1 - OSPF external type 1, E2 - OSPF external type 2
       i - IS-IS, su - IS-IS summary, L1 - IS-IS level-1, L2 - IS-IS level-2
       ia - IS-IS inter area, * - candidate default, U - per-user static route
       o - ODR, P - periodic downloaded static route

Gateway of last resort is not set

C    192.168.0.0/24 is directly connected, FastEthernet0/0.10
C    192.168.1.0/24 is directly connected, FastEthernet0/0.20
```

图 9-7　路由器 R1 的路由表

步骤 5：在计算机 PC1 的命令行界面输入 ping 192.168.1.1，检验联通性，如图 9-8 所示。

```
C:\>ping 192.168.1.1

正在 Ping 192.168.1.1 具有 32 字节的数据：
来自 192.168.1.1 的回复: 字节=32 时间<1ms TTL=127
来自 192.168.1.1 的回复: 字节=32 时间<1ms TTL=127
来自 192.168.1.1 的回复: 字节=32 时间<1ms TTL=127
来自 192.168.1.1 的回复: 字节=32 时间<1ms TTL=127

192.168.1.1 的 Ping 统计信息:
    数据包: 已发送 = 4，已接收 = 4，丢失 = 0 (0% 丢失),
往返行程的估计时间(以毫秒为单位):
    最短 = 0ms，最长 = 0ms，平均 = 0ms
```

图 9-8　从计算机 PC1 ping 计算机 PC2

本小节将路由器 R1 的 f0/0 口连接交换机 S1 的 f0/24 口，然后在路由器 R1 上配置子接口，每个子接口都传递一个 VLAN 的流量。路由器 R1 和交换机 S1 互联的链路要传递多个 VLAN 的流量，将交换机 S1 的 f0/24 口配置为 trunk 模式。通过查看路由表可以发现，当路由器的子接口配置完成后，路由表中出现了对应子接口的直连路由。通过检验网络的联通性可以发现，计算机 PC1 和 PC2 实现了互通。

9.5　项目小结

本项目完成了传统方式 VLAN 间路由的配置和单臂路由的配置。要实现位于不同 VLAN 的计算机互通，可以采用传统方式的 VLAN 间路由或者单臂路由的方式。传统方式的 VLAN 间路由需要路由器和交换机有较多的物理接口，单臂路由避免了这个问题。但是子接口所在的物理接口承载了所有通过该接口传递的 VLAN 的流量，一旦该物理接口的带宽不足以满足所有 VLAN 同时满负荷传输的需求，这个接口就有可能成为网络的瓶颈。

9.6　拓展训练

本项目的拓展训练网络拓扑如图 9-9 所示，要求完成如下配置：

（1）根据图 9-9 所示网络拓扑中的标注，为交换机 S1 和 S2 创建 VLAN，并把端口分配给 VLAN；
（2）完成 VLAN 中继的配置，实现计算机 PC1 和 PC3 互通、计算机 PC2 和 PC4 互通；
（3）完成单臂路由的配置，实现计算机 PC1、PC2、PC3 和 PC4 互通。

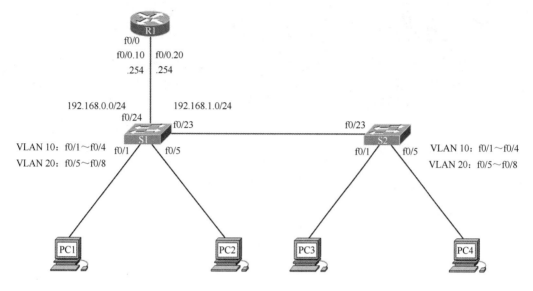

图 9-9　项目 9 拓展训练网络拓扑

项目 10

三层交换机VLAN间路由的配置

10.1 用户需求

某学校网络拓扑如图 10-1 所示。两台计算机位于不同的 VLAN，怎样用三层交换机实现位于不同 VLAN 的计算机互通？

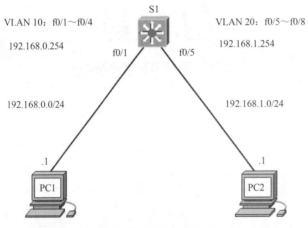

图 10-1 某学校网络拓扑

10.2 知识梳理

10.2.1 三层交换机

二层交换机只根据开放系统互联（Open System Interconnect，OSI）参考模型第二层（数据链路层）的 MAC 地址执行交换和过滤操作，建立 MAC 地址表，使用 MAC 地址表做出转发决策，对网络协议和用户应用程序完全透明。

三层交换机不仅可以使用第二层的 MAC 地址来做出转发决策，还可以使用 IP 地址。三层交换机不仅知道哪些 MAC 地址与交换机的物理接口关联，而且还知道哪些 IP 地址与交换机的物理接口关联。三层交换机还可以通过检查以太网数据包中的第三层信息来作出转发决策，可以像路由器一样在不同的 LAN 网段之间转发数据包，但是不能完全取代路由器。使用三层交换机的根本目的是加快大型局域网内部的数据交换,能够做到"一次路由,多次转发",数据包转发等规律性的过程由硬件实现,而路由信息更新、路由表维护、路由计算、路由确定等功能由软件实现。

10.2.2 三层交换机端口

三层交换机端口主要分为二层口(交换口)和三层口(路由口)两种。

1. 二层口

三层交换机的二层口与普通二层交换机的端口一样,不能为其配置 IP 地址。默认情况下,思科三层交换机的所有端口都是工作在二层的,只能识别二层数据帧。

2. 三层口

三层交换机的三层口与路由器的接口一样,不与特定 VLAN 相关,需要为其配置 IP 地址,思科交换机的三层口不支持子接口。

10.2.3 三层交换机的 VLAN 间路由

三层交换机的 VLAN 间路由是通过 SVI 来实现的。SVI 是配置在多层交换机中的虚拟接口。作为 VLAN 内主机的网关,可以为交换机上的任何 VLAN 创建 SVI。SVI 是路由口,VLAN 的 SVI 为与该 VLAN 相关联的所有交换机之间传输的数据包提供了第三层处理过程。

三层交换机的 VLAN 间路由

10.2.4 配置命令

1. 配置交换机的端口为三层口

```
Switch(config)#interface interface-id
Switch(config-if)#no switchport
```

2. 开启三层交换机的路由功能

```
Switch(config)#ip routing
```

3. 创建 VLAN 的虚拟接口并配置 IP 地址

```
Switch(config)#interface vlan vlan-id
Switch(config-if)#ip address ip-address mask
```

vlan-id:要创建的 SVI 的 VLAN ID。
ip-address:SVI 的 IP 地址。
mask:SVI 的子网掩码。

10.3 方案设计

在图 10-1 所示的网络拓扑中,要实现位于不同 VLAN 的计算机互通,可以使用三层交换机。三层交换机既可以完成 VLAN 的划分,又可以实现 VLAN 间路由,相当于把路由器和普通二层交换机合并。

10.4 项目实施

本节网络拓扑如图 10-1 所示,计算机的 IP 地址已经配置完成,要求完成三层交换机 VLAN 间路由的配置，实现计算机 PC1 和 PC2 互通。

步骤 1：在交换机 S1 的全局配置模式下配置 VLAN，输入以下代码。

```
S1(config)#vlan 10
S1(config-vlan)#vlan 20
S1(config-vlan)#interface range f0/1-4
S1(config-if-range)#switchport mode access
S1(config-if-range)#switchport access vlan 10
S1(config-if-range)#interface range f0/5-8
S1(config-if-range)#switchport mode access
S1(config-if-range)#switchport access vlan 20
```

步骤 2：在交换机 S1 的全局配置模式下配置 SVI，输入以下代码。

```
S1(config)#interface vlan 10
S1(config-if)#ip address 192.168.0.254 255.255.255.0
S1(config-if)#no shutdown
S1(config-if)#interface vlan 20
S1(config-if)#ip add 192.168.1.254 255.255.255.0
S1(config-if)#no shutdown
```

步骤 3：在交换机 S1 的全局配置模式下开启三层交换机的路由功能，输入以下代码。

```
S1(config)#ip routing
```

步骤 4：在交换机 S1 的特权执行模式下输入 show ip route，查看三层交换机 S1 的路由表，如图 10-2 所示。

```
S1#show ip route
Codes: L - local, C - connected, S - static, R - RIP, M - mobile, B - BGP
       D - EIGRP, EX - EIGRP external, O - OSPF, IA - OSPF inter area
       N1 - OSPF NSSA external type 1, N2 - OSPF NSSA external type 2
       E1 - OSPF external type 1, E2 - OSPF external type 2
       i - IS-IS, su - IS-IS summary, L1 - IS-IS level-1, L2 - IS-IS level-2
       ia - IS-IS inter area, * - candidate default, U - per-user static route
       o - ODR, P - periodic downloaded static route, H - NHRP, l - LISP
       + - replicated route, % - next hop override

Gateway of last resort is not set

     192.168.0.0/24 is variably subnetted, 2 subnets, 2 masks
C       192.168.0.0/24 is directly connected, Vlan10
L       192.168.0.254/32 is directly connected, Vlan10
     192.168.1.0/24 is variably subnetted, 2 subnets, 2 masks
C       192.168.1.0/24 is directly connected, Vlan20
L       192.168.1.254/32 is directly connected, Vlan20
```

图 10-2　三层交换机 S1 的路由表

步骤 5：在计算机 PC1 的命令行界面输入 ping 192.168.1.1，检验联通性，如图 10-3 所示。

```
C:\>ping 192.168.1.1

正在 Ping 192.168.1.1 具有 32 字节的数据:
来自 192.168.1.1 的回复: 字节=32 时间<1ms TTL=127
来自 192.168.1.1 的回复: 字节=32 时间<1ms TTL=127
来自 192.168.1.1 的回复: 字节=32 时间<1ms TTL=127
来自 192.168.1.1 的回复: 字节=32 时间<1ms TTL=127

192.168.1.1 的 Ping 统计信息:
    数据包: 已发送 = 4，已接收 = 4，丢失 = 0 (0% 丢失)，
往返行程的估计时间(以毫秒为单位):
    最短 = 0ms，最长 = 0ms，平均 = 0ms
```

图 10-3　从计算机 PC1 ping 计算机 PC2

本节先在三层交换机 S1 上配置 VLAN，通过配置交换机的 SVI，实现三层交换机 VLAN 间路由，然后启用三层交换机的路由功能。通过查看路由表可以发现，当完成了 SVI 的配置后，三层交换机的路由表中会出现对应的直连路由。通过检验联通性可以发现，网络实现了互通。

10.5 项目小结

本项目完成了三层交换机 VLAN 间路由的配置，三层交换机有二层交换功能和三层路由功能，可将三层交换机的端口配置为二层交换口或三层路由口。为了实现三层交换机 VLAN 间路由，本项目在三层交换机上配置了 VLAN，同时配置了 SVI，提供了各 VLAN 之间的路由功能，以便属于不同 VLAN 的主机能够相互通信。

10.6 拓展训练

本项目的拓展训练网络拓扑如图 10-4 所示，要求完成如下配置：
（1）根据图 10-4 所示网络拓扑中的标注，为交换机 S1 和 S2 创建 VLAN，并把端口分配给 VLAN；
（2）完成 VLAN 中继的配置，实现计算机 PC1 与 PC3 互通，计算机 PC2 与 PC4 互通；
（3）完成三层交换机 VLAN 间路由的配置，实现计算机 PC1、PC2、PC3 和 PC4 互通。

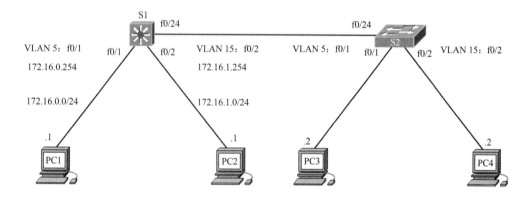

图 10-4　项目 10 拓展训练网络拓扑

模块三

网络访问控制的配置

网络访问控制可以保护网络资源不被非法使用和访问,访问控制列表(ACL)可以限制网络的访问范围、过滤网络中的流量,是控制访问的一种网络技术手段。本模块主要介绍部署 ACL 限制网络访问范围和部署 ACL 限制网络流量,完成网络访问范围限制和网络流量限制的配置。

项目 11
部署ACL限制网络访问范围

11.1 用户需求

某学校网络拓扑如图 11-1 所示，计算机 PC1 位于办公网络，计算机 PC2 位于食堂网络，计算机 PC3 位于学生网络，怎样做可以让学生网络中的计算机无法访问办公网络？

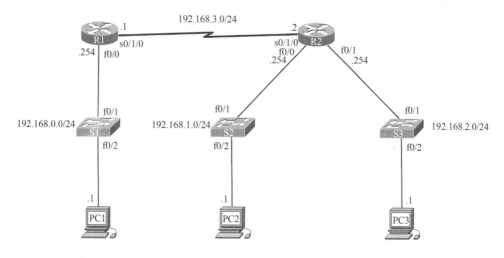

图 11-1 某学校网络拓扑

11.2 知识梳理

11.2.1 ACL 的概念

访问控制列表（Access Control List，ACL）是一种路由器配置脚本，它根据从报头中发现的条件来控制路由器允许或拒绝报文通过。ACL 在控制进出网络的流量方面非常有用。

ACL 的概念

11.2.2 ACL 的用途

ACL 的用途主要有流量过滤、流量分类和允许或拒绝对网络服务的访问。

1. 流量过滤

ACL 可以根据条件语句对数据包进行匹配，在保证合法用户的报文通过的同时拒绝非法用户的访问。例如，允许办公网络中的设备访问办公网络、学生网络和食堂网络，不允许学生网络中的设备访问办公网络。

2. 流量分类

路由器在发送与接收路由信息时，ACL 可以用于匹配路由信息的目的地址，从而对流量进行分类，过滤掉不需要的路由信息。

3. 允许或拒绝对网络服务的访问

ACL 可以根据特定服务的端口号进行匹配，可以允许或拒绝用户访问特定网络服务。

11.2.3 ACL 的工作原理

当每个数据包经过关联 ACL 的接口时，都会与 ACL 中的语句从上到下逐条进行对比，以便发现符合条件的数据包。ACL 使用允许或拒绝规则来决定是否转发数据包。默认情况下，路由器上没有配置任何 ACL，不会过滤流量，所有可以被路由器路由的数据包都会经过路由器到达下一个网段。

1. 入站 ACL 的工作原理

对于启用了入站 ACL 的接口，如果有数据包传入，则由入站 ACL 进行处理，将其从上到下逐条与 ACL 中的语句进行对比。当数据包匹配到 ACL 中的某条语句时，如果 ACL 的操作是 permit，那么数据包会被允许通过，路由器会通过查找路由表把数据包转发出去；如果 ACL 的操作是 deny，那么数据包会被丢弃。ACL 中的最后一条语句是隐式拒绝语句，如果匹配不到 ACL 中的语句，那么数据包会被丢弃。

2. 出站 ACL 的工作原理

对于启用了出站 ACL 的接口，当数据包路由到该接口时，由出站 ACL 进行处理，将其从上到下逐条与 ACL 中的语句进行对比。当数据包匹配到 ACL 中的某条语句时，如果 ACL 的操作是 permit，那么数据包会被允许从该接口转发出去；如果 ACL 的操作是 deny，那么数据包会被丢弃。ACL 中的最后一条语句是隐式拒绝语句，所以如果匹配不到 ACL 中的语句，那么数据包会被丢弃。

11.2.4 ACL 的类型

ACL 的类型主要分成两种，分别是标准 ACL 和扩展 ACL。根据配置方式，标准 ACL 可分为标准命名 ACL 和标准编号 ACL。

1. 标准 ACL

标准 ACL 根据源 IP 地址过滤数据包。数据包中包含的目的地址和端口号无关紧要。

2. 扩展 ACL

扩展 ACL 根据多种属性（如协议类型、源 IP 地址、目的 IP 地址、源 TCP 或 UDP 端口号、目的 TCP 或 UDP 端口号）过滤数据包，并依据协议类型进行更为精确的控制。

通配符掩码和通配符掩码关键字

11.2.5 通配符掩码

通配符掩码是 32 位二进制数，使用二进制数 1 和 0 过滤单个 IP 地址或一组 IP 地址，用于确定应该为地址匹配多少位源 IP 地址或目的 IP 地址。通配符掩码位为 0 表示精确匹配地址中对应位的值；通配符掩码位为 1 表示忽略地址中对应位的值。

11.2.6 通配符掩码关键字

通配符掩码关键字 host 和 any 可以用来标识常用的通配符掩码，使用这些关键字避免了在标识特定主机或网络时输入通配符掩码带来的麻烦。

host 关键字可替代 0.0.0.0 通配符掩码。此通配符掩码表示必须匹配所有 IP 地址位，即仅匹配一台主机。

any 关键字可替代 IP 地址和 255.255.255.255 通配符掩码。该通配符掩码表示允许任何地址的数据通过，即匹配任何地址。

11.2.7 ACL 创建原则

1. 每种协议一个 ACL

要控制接口上的流量，必须为接口上启用的每种协议创建相应的 ACL。

2. 每个方向一个 ACL

一个 ACL 只能控制接口上一个方向的流量。因此，要控制入站流量和出站流量，必须分别定义两个 ACL。

3. 每个接口一个 ACL

一个 ACL 只能控制一个接口上的流量，必须为每个接口定义相应的 ACL。

11.2.8 标准 ACL 的放置位置

ACL 的放置位置决定了是否能有效减少不必要的流量，在适当的位置放置 ACL 可以过滤掉不必要的流量，使网络更加高效。因为标准 ACL 不会指定目的地址，所以标准 ACL 的放置位置应该尽可能靠近目的地。

11.2.9 配置命令

1. 配置标准编号 ACL

```
Router(config)#access-list access-list-number {deny|permit|remark}
source [source-wildcard] [log]
```

access-list-number：ACL 的编号。这是一个十进制数，适用于编号范围为 1~99 或 1300~1999 的标准 ACL。

deny：匹配条件语句时拒绝访问数据包。

permit：匹配条件语句时允许访问数据包。

remark：在 IP ACL 中添加备注，提高列表的可读性。

source：发送的源数据包的网络地址或主机地址。

source-wildcard：可选参数，对应源应用的通配符掩码。

log：可选参数，为匹配条目的数据包生成信息性日志消息，该消息随后将被发送到 Console 口。消息内容包括 ACL 编号、数据包被允许或被拒绝、源地址及数据包数目。此消息在出现与条件语句匹配的第一个数据包时生成，随后每 5min 生成一次，消息内容会包含过去的 5min 内允许或拒绝的数据包的数量。

2. 配置标准命名 ACL

```
Router(config)#ip access-list standard name
Router(config-std-nacl)#{deny|permit|remark} source [source-wildcard] [log]
```

name：ACL 的名称，名称中可以包含字母、数字，不能包含空格或标点，而且必须以字母开头。

3. 在接口上应用 ACL

```
Router(config)#interface type number
Router(config-if)# ip access-group {access-list-number|name} {in|out}
```

type number：接口的类型和编号。

access-list-number：ACL 的编号。

name：ACL 的名称。

in：数据流入路由器接口时使用的参数。

out：数据流出路由器接口时使用的参数。

4. 查看 ACL

```
Router#show access-lists
```

5. 查看接口的配置

```
Router#show ip interface
```

该命令可以查看接口应用的 ACL。

6. 清除 ACL 的统计信息

```
Router#clear access-list counters 1
```

该命令可以单独使用，也可以与特定的 ACL 的编号或者名称一起使用。该命令的功能是清除 ACL 的统计信息。

11.2.10 ACL 的编辑

1. 使用文本编辑器

可以使用文本编辑器（如 Microsoft 记事本）创建或编辑 ACL，然后将 ACL 复制并粘贴到路由器中。对于现有的 ACL，可以使用 show running-config 命令显示 ACL，然后将现有的 ACL 复制并粘贴到文本编辑器中，进行必要的更改后再将其复制并粘贴到路由器中。

应该注意的是，在修改 ACL 之前要将有关的 ACL 从接口删除。如果将已删除的 ACL 应用于接口，那么某些版本的 IOS 会拒绝所有流量。

2. 使用序列号

使用 show access-lists 命令显示当前 ACL，执行此命令后输出的每条 ACL 语句都会显示序列号（在输入 ACL 语句时，会自动为其分配序列号），然后输入用于配置标准命名 ACL 的 ip access-list standard *name* 命令。如果修改的是标准编号 ACL，则将 ACL 的编号作为 ACL 的名称；如果修改的是标准命名 ACL，则直接输入 ACL 的名称。把需要修改的语句删除后再添加新语句，其中，序列号可以控制添加语句的位置。

 注意 使用与现有语句相同的序列号并不能覆盖现有语句。必须先删除现有语句，然后才能添加新语句。

11.3 方案设计

在图 11-1 所示的网络拓扑中，要实现学生网络中的计算机无法访问办公网络，可以采用标准 ACL 拒绝来自学生网络中的数据。标准 ACL 依据数据包的源地址进行控制，配置 ACL 语句时要使用学生网络的网络地址和通配符掩码，标准 ACL 的放置位置要尽量靠近目的地，所以，要在路由器 R1 上创建 ACL，在路由器 R1 的 f0/0 接口上应用 ACL。

11.4 项目实施

11.4.1 标准编号 ACL 的配置

本小节网络拓扑如图 11-1 所示，计算机 PC1 位于办公网络，计算机 PC2 位于食堂网络，计算机 PC3 位于学生网络。路由器的接口和计算机的 IP 地址已经配置完成，要求完成 RIPv2 和标准编号 ACL 的配置，实现学生网络中的计算机无法访问办公网络。

步骤 1：在路由器 R1 的全局配置模式下配置 RIPv2，输入以下代码。

```
R1(config)#router rip
R1(config-router)#version 2
R1(config-router)#network 192.168.0.0
R1(config-router)#network 192.168.3.0
R1(config-router)#no auto-summary
```

步骤 2：在路由器 R2 的全局配置模式下配置 RIPv2，输入以下代码。

```
R2(config)#router rip
R2(config-router)#version 2
R2(config-router)#network 192.168.1.0
R2(config-router)#network 192.168.2.0
R2(config-router)#network 192.168.3.0
R2(config-router)#no auto-summary
```

步骤 3：在路由器 R1 的特权执行模式下输入 show ip route，查看路由表，如图 11-2 所示。

```
R1#show ip route
Codes: C - connected, S - static, R - RIP, M - mobile, B - BGP
       D - EIGRP, EX - EIGRP external, O - OSPF, IA - OSPF inter area
       N1 - OSPF NSSA external type 1, N2 - OSPF NSSA external type 2
       E1 - OSPF external type 1, E2 - OSPF external type 2
       i - IS-IS, su - IS-IS summary, L1 - IS-IS level-1, L2 - IS-IS level-2
       ia - IS-IS inter area, * - candidate default, U - per-user static route
       o - ODR, P - periodic downloaded static route

Gateway of last resort is not set

C    192.168.0.0/24 is directly connected, FastEthernet0/0
R    192.168.1.0/24 [120/1] via 192.168.3.2, 00:00:12, Serial0/1/0
R    192.168.2.0/24 [120/1] via 192.168.3.2, 00:00:12, Serial0/1/0
C    192.168.3.0/24 is directly connected, Serial0/1/0
```

图 11-2　路由器 R1 的路由表

步骤 4：在路由器 R2 的特权执行模式下输入 show ip route，查看路由表，如图 11-3 所示。

```
R2#show ip route
Codes: C - connected, S - static, R - RIP, M - mobile, B - BGP
       D - EIGRP, EX - EIGRP external, O - OSPF, IA - OSPF inter area
       N1 - OSPF NSSA external type 1, N2 - OSPF NSSA external type 2
       E1 - OSPF external type 1, E2 - OSPF external type 2
       i - IS-IS, su - IS-IS summary, L1 - IS-IS level-1, L2 - IS-IS level-2
       ia - IS-IS inter area, * - candidate default, U - per-user static route
       o - ODR, P - periodic downloaded static route

Gateway of last resort is not set

R    192.168.0.0/24 [120/1] via 192.168.3.1, 00:00:23, Serial0/1/0
C    192.168.1.0/24 is directly connected, FastEthernet0/0
C    192.168.2.0/24 is directly connected, FastEthernet0/1
C    192.168.3.0/24 is directly connected, Serial0/1/0
```

图 11-3　路由器 R2 的路由表

步骤 5：在计算机 PC3 的命令行界面输入 ping 192.168.0.1，检验联通性，如图 11-4 所示。

```
C:\>ping 192.168.0.1

正在 Ping 192.168.0.1 具有 32 字节的数据:
来自 192.168.0.1 的回复: 字节=32 时间=9ms TTL=126
来自 192.168.0.1 的回复: 字节=32 时间=9ms TTL=126
来自 192.168.0.1 的回复: 字节=32 时间=9ms TTL=126
来自 192.168.0.1 的回复: 字节=32 时间=9ms TTL=126

192.168.0.1 的 Ping 统计信息:
    数据包: 已发送 = 4, 已接收 = 4, 丢失 = 0 (0% 丢失),
往返行程的估计时间(以毫秒为单位):
    最短 = 9ms, 最长 = 9ms, 平均 = 9ms
```

图 11-4　从计算机 PC3 ping 计算机 PC1

步骤 6：在计算机 PC3 的命令行界面输入 ping 192.168.1.1，检验联通性，如图 11-5 所示。

```
C:\>ping 192.168.1.1

正在 Ping 192.168.1.1 具有 32 字节的数据:
来自 192.168.1.1 的回复: 字节=32 时间<1ms TTL=127
来自 192.168.1.1 的回复: 字节=32 时间<1ms TTL=127
来自 192.168.1.1 的回复: 字节=32 时间<1ms TTL=127
来自 192.168.1.1 的回复: 字节=32 时间<1ms TTL=127

192.168.1.1 的 Ping 统计信息:
    数据包: 已发送 = 4, 已接收 = 4, 丢失 = 0 (0% 丢失),
往返行程的估计时间(以毫秒为单位):
    最短 = 0ms, 最长 = 0ms, 平均 = 0ms
```

图 11-5　从计算机 PC3 ping 计算机 PC2

步骤 7：在路由器 R1 的全局配置模式下配置 ACL，输入以下代码。

```
R1(config)#access-list 1 deny 192.168.2.0 0.0.0.255
R1(config)#access-list 1 permit any
```

步骤 8：在路由器 R1 的全局配置模式下，在接口上应用 ACL，输入以下代码。

```
R1(config)#interface f0/0
R1(config-if)#ip access-group 1 out
```

步骤 9：在计算机 PC3 的命令行界面输入 ping 192.168.0.1，检验联通性，如图 11-6 所示。

```
C:\>ping 192.168.0.1
正在 Ping 192.168.0.1 具有 32 字节的数据:
来自 192.168.3.1 的回复: 无法访问目标网。
来自 192.168.3.1 的回复: 无法访问目标网。
来自 192.168.3.1 的回复: 无法访问目标网。
来自 192.168.3.1 的回复: 无法访问目标网。

192.168.0.1 的 Ping 统计信息:
    数据包: 已发送 = 4, 已接收 = 4, 丢失 = 0 (0% 丢失),
```

图 11-6　从计算机 PC3 ping 计算机 PC1

步骤 10：在计算机 PC3 的命令行界面输入 ping 192.168.1.1，检验联通性，如图 11-7 所示。

```
C:\>ping 192.168.1.1
正在 Ping 192.168.1.1 具有 32 字节的数据:
来自 192.168.1.1 的回复: 字节=32 时间<1ms TTL=127
来自 192.168.1.1 的回复: 字节=32 时间<1ms TTL=127
来自 192.168.1.1 的回复: 字节=32 时间<1ms TTL=127
来自 192.168.1.1 的回复: 字节=32 时间<1ms TTL=127

192.168.1.1 的 Ping 统计信息:
    数据包: 已发送 = 4, 已接收 = 4, 丢失 = 0 (0% 丢失),
往返行程的估计时间(以毫秒为单位):
    最短 = 0ms, 最长 = 0ms, 平均 = 0ms
```

图 11-7　从计算机 PC3 ping 计算机 PC2

通过查看路由表和检验联通性可以发现，在配置 ACL 之前，路由表中的路由完整，计算机 PC1、PC2 和 PC3 能够互通，当配置了 ACL 后，位于学生网络中的计算机 PC3 无法访问办公网络中的计算机 PC1。

11.4.2　标准命名 ACL 的配置

本小节网络拓扑如图 11-8 所示，计算机 PC1 位于办公网络，计算机 PC2 位于食堂网络，计算机 PC3 和 PC4 位于学生网络。路由器的接口和计算机的 IP 地址已经配置完成，已经实现了全网互通，要求完成标准命名 ACL 的配置，实现学生网络中除了计算机 PC4 外的其他设备无法访问办公网络。

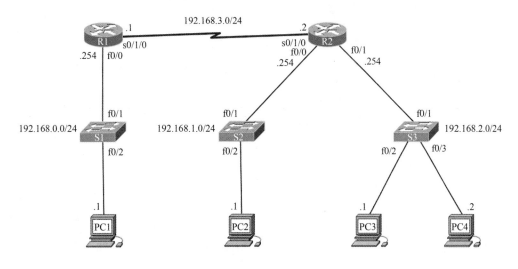

图 11-8　网络拓扑

步骤 1：在路由器 R1 的全局配置模式下配置 ACL，输入以下代码。

```
R1(config)# ip access-list standard denypc3
R1(config-std-nacl)#permit 192.168.2.2 0.0.0.0
R1(config-std-nacl)#deny 192.168.2.0 0.0.0.255
R1(config-std-nacl)#permit any
```

当 ACL 语句中的条件是某个地址时，可以用 host 关键字，如第 2 条语句也可以写成：

```
R1(config-std-nacl)#permit host 192.168.2.2
```

步骤 2：在路由器 R1 的全局配置模式下，在接口上应用 ACL，输入以下代码。

```
R1(config)#interface f0/0
R1(config-if)#ip access-group denypc3 out
```

步骤 3：在计算机 PC3 的命令行界面输入 ping 192.168.0.1，检验联通性，如图 11-9 所示。

```
C:\>ping 192.168.0.1

正在 Ping 192.168.0.1 具有 32 字节的数据：
来自 192.168.3.1 的回复：无法访问目标网。
来自 192.168.3.1 的回复：无法访问目标网。
来自 192.168.3.1 的回复：无法访问目标网。
来自 192.168.3.1 的回复：无法访问目标网。

192.168.0.1 的 Ping 统计信息：
    数据包：已发送 = 4，已接收 = 4，丢失 = 0 (0% 丢失)，
```

图 11-9　从计算机 PC3 ping 计算机 PC1

步骤 4：在计算机 PC4 的命令行界面输入 ping 192.168.0.1，检验联通性，如图 11-10 所示。

```
C:\>ping 192.168.0.1

正在 Ping 192.168.0.1 具有 32 字节的数据：
来自 192.168.0.1 的回复：字节=32 时间=9ms TTL=126
来自 192.168.0.1 的回复：字节=32 时间=9ms TTL=126
来自 192.168.0.1 的回复：字节=32 时间=9ms TTL=126
来自 192.168.0.1 的回复：字节=32 时间=9ms TTL=126

192.168.0.1 的 Ping 统计信息：
    数据包：已发送 = 4，已接收 = 4，丢失 = 0 (0% 丢失)，
往返行程的估计时间(以毫秒为单位)：
    最短 = 9ms，最长 = 9ms，平均 = 9ms
```

图 11-10　从计算机 PC4 ping 计算机 PC1

步骤 5：修改步骤 1 中的配置命令，把第 2 条和第 3 条语句互换顺序，写成：

```
R1(config)# ip access-list standard denypc3
R1(config-std-nacl)#deny 192.168.2.0 0.0.0.255
R1(config-std-nacl)#permit 192.168.2.2 0.0.0.0
R1(config-std-nacl)#permit any
```

步骤 6：在计算机 PC3 的命令行界面输入 ping 192.168.0.1，检验联通性，如图 11-11 所示。

```
C:\>ping 192.168.0.1

正在 Ping 192.168.0.1 具有 32 字节的数据:
来自 192.168.3.1 的回复: 无法访问目标网。
来自 192.168.3.1 的回复: 无法访问目标网。
来自 192.168.3.1 的回复: 无法访问目标网。
来自 192.168.3.1 的回复: 无法访问目标网。

192.168.0.1 的 Ping 统计信息:
    数据包: 已发送 = 4, 已接收 = 4, 丢失 = 0 (0% 丢失),
```

图 11-11　从计算机 PC3 ping 计算机 PC1

步骤 7：在计算机 PC4 的命令行界面输入 ping 192.168.0.1，检验联通性，如图 11-12 所示。

```
C:\>ping 192.168.0.1

正在 Ping 192.168.0.1 具有 32 字节的数据:
来自 192.168.3.1 的回复: 无法访问目标网。
来自 192.168.3.1 的回复: 无法访问目标网。
来自 192.168.3.1 的回复: 无法访问目标网。
来自 192.168.3.1 的回复: 无法访问目标网。

192.168.0.1 的 Ping 统计信息:
    数据包: 已发送 = 4, 已接收 = 4, 丢失 = 0 (0% 丢失),
```

图 11-12　从计算机 PC4 ping 计算机 PC1

通过检验联通性可以发现，在完成了 ACL 的配置后，学生网络中的计算机 PC4 可以访问办公网络中的计算机 PC1，而学生网络中的其他计算机，即 PC3 无法访问办公网络中的计算机 PC1。当把步骤 1 中配置 ACL 的第 2 条和第 3 条语句的顺序互换后，学生网络中的计算机 PC3 和 PC4 都无法访问办公网络中的计算机 PC1 了。ACL 在执行访问控制的时候，从第 1 条语句开始依次对比，若匹配到第 1 条语句，就会按照第 1 条语句的操作来处理，不会再对比第 2 条语句，这样来自 192.168.2.0 网络中的所有数据都将被拒绝。所以，ACL 语句的顺序非常重要，在配置 ACL 时不要随意调换语句的顺序，语句的顺序会影响 ACL 的访问控制结果。

11.5　项目小结

本项目完成了部署 ACL 限制网络访问范围。标准 ACL 根据数据包的源地址对数据包进行访问控制，标准 ACL 包括标准命名 ACL 和标准编号 ACL。ACL 中的最后一条语句是隐式拒绝语句，所以在配置 ACL 时必须至少有一条 permit 语句，否则应用 ACL 的接口会拒绝所有数据。在配置 ACL 时要注意 ACL 语句的顺序，语句内容相同但顺序不同的 ACL 可能产生不同的访问控制结果。

11.6　拓展训练

本项目的拓展训练网络拓扑如图 11-13 所示，要求完成如下配置：
（1）完成计算机 IP 地址、路由器接口和路由的配置，实现计算机 PC1、PC2 和 PC3 互通；

（2）配置标准编号 ACL，实现计算机 PC1 无法访问计算机 PC2。

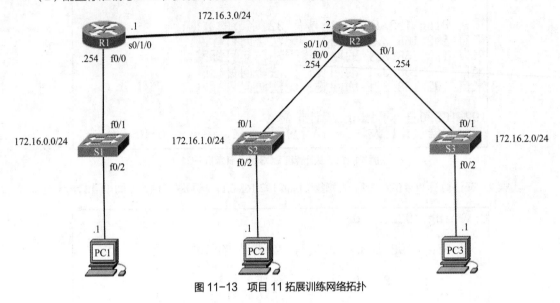

图 11-13　项目 11 拓展训练网络拓扑

项目 12
部署ACL限制网络流量

12.1 用户需求

某学校网络拓扑如图 12-1 所示,怎样实现 192.168.1.0/24 网络中的设备无法远程登录路由器 R2?

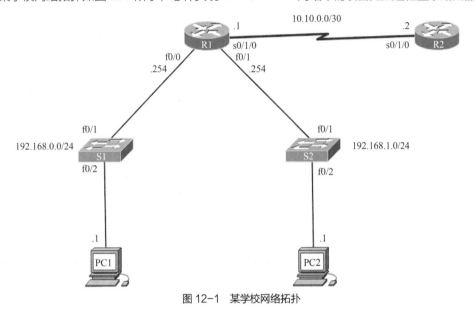

图 12-1 某学校网络拓扑

12.2 知识梳理

12.2.1 扩展 ACL

扩展 ACL

扩展 ACL 根据多种属性过滤 IP 数据包,可以更加精确地控制流量,过滤控制范围更广,可以提升安全性。扩展 ACL 可以根据数据包源 IP 地址、目的 IP 地址、协议类型、源及目的 TCP 和 UDP 端口号等对数据进行访问控制。扩展 ACL 的编号范围为 100~199 和 2000~2699。

根据配置方式,扩展 ACL 可分为扩展编号 ACL 和扩展命名 ACL。

12.2.2 端口号

端口号是 TCP 和 UDP 中标识应用程序的唯一报头字段的标识符。支持 TCP 和 UDP 的服务将对正在通信的不同应用程序进行跟踪,对每个应用程序的数据段和数据报文用端口号进行标识。在每个数据段或数据报文的报头内,各有一个源端口号和目的端口号。源端口号是与本地主机上的始发应用程序相关联的通信端口号,目的端口号是与远程主机上的目的应用程序相关联的通信端口号。端口号有如下类型。

1. 公认端口

公认端口(端口号: 0~1023)用于特定的服务或应用程序。通过为服务或应用程序定义公认端口,客户端应用程序可以向服务器请求特定端口,实现与服务器相关服务的连接。常用的公认端口如表 12-1 所示。

表 12-1 常用的公认端口

TCP/UDP 端口	端口号	服务或应用程序
TCP 端口	21	文件传输协议(File Transfer Protocol,FTP)服务
TCP 端口	23	telnet 服务
TCP 端口	25	简单邮件传送协议(Simple Mail Transfer Protocol,SMTP)服务
TCP 端口	80	HTTP 服务
TCP 端口	143	交互式邮件存取协议(Internet Mail Access Protocol,IMAP)服务
TCP 端口	194	Internet 中继聊天(IRC)
TCP 端口	443	超文本传输协议安全(Hypertext Transfer Protocol Secure,HTTPS)服务
UDP 端口	69	简单文件传送协议(Trivial File Transfer Protocol,TFTP)服务
UDP 端口	520	RIP 服务
TCP/UDP 端口	53	域名系统(Domain Name System,DNS)协议服务
TCP/UDP 端口	161	SNMP 服务
TCP/UDP 端口	531	AOL instant Messenger、IRC

2. 已注册端口

已注册端口(端口号: 1024~49151)将被分配给用户进程或应用程序。这些端口主要用于用户选择安装的应用程序,与已经分配了公认端口的常用应用程序有所区别。这些端口在没有被服务器资源占用时,可被客户端动态选用为源端口。

3. 动态或私有端口

动态或私有端口(端口号: 49152~65535)也称为临时端口,动态或私有端口往往在开始连接时被动态分配给客户端应用程序。客户端一般很少使用动态或私有端口连接服务,只有一些点对点文件共享程序使用这类端口。

12.2.3 扩展 ACL 的放置位置

扩展 ACL 应尽量靠近网络中数据流量的起始点,以便在流量进入网络之前及时过滤掉不需要的流量。

12.2.4 基于时间的 ACL

基于时间的 ACL 允许管理员根据时间对网络资源执行访问控制。要使用基于时间的 ACL,需要创建时间范围，指定一周或一天内的时段。时间范围可被命名并应用到相应功能上，以控制其特定时段内的访问权限。

12.2.5 配置命令

1. 配置扩展编号 ACL

```
Router(config)#access-list access-list-number {deny|permit|remark} protocol
source [source-wildcard] [operator operand] [port port-number or name] destination
[destination-wildcard] [operator operand] [port port-number or name] [established]
```

access-list-number：扩展 ACL 的编号，使用范围为 100~199 或 2000~2699 的数字标识扩展 ACL。

deny：匹配条件语句时拒绝访问。

permit：匹配条件语句时允许访问。

remark：在 IP ACL 中添加备注，提高列表的可读性。

protocol：协议的名称或编号，可选的关键字包括 icmp、ip、tcp 或 udp 等。若要匹配所有互联网（Internet）协议，则使用 ip 关键字。

source：发送数据包（数据包的源）的网络地址或主机地址。

source-wildcard：可选参数，对应源应用的通配符掩码。

destination：数据包要发往的目的网络地址或主机地址。

destination-wildcard：对应目的地址应用的通配符掩码。

operator：可选参数，用于指定 ACL 中对源或目的端口的比较操作符，可用的操作符包括 lt（小于）、gt（大于）、eq（等于）、neq（不等于）和 range（范围）。

port：可选参数，TCP 或 UDP 端口的十进制编号或名称。

established：可选参数，仅用于 TCP，指示已建立的连接。该参数允许源网络的流量返回该网络。如果没有该参数，那么客户端可以向服务器发送流量，但是不会收到从服务器返回的流量。

2. 配置扩展命名 ACL

```
Router(config)#ip access-list extended name
Router(config-std-nacl)#{deny|permit|remark} protocol source [source-wildcard]
[operator operand] [port port-number or name] destination [destination-wildcard]
[operator operand] [port port-number or name] [time-range time-range-name]
[established]
```

name：ACL 的名称，名称中可以包含字母、数字，不能包含空格或标点，而且必须以字母开头。

time-range-name：过滤数据包的时间范围的名称。

3. 在接口上应用 ACL

```
Router(config)#interface type number
Router(config-if)# ip access-group {access-list-number|name} {in|out}
```

type number：接口的类型和编号。

access-list-number：ACL 的编号。

name：ACL 的名称。

in：数据流入路由器接口时使用的参数。

out：数据流出路由器接口时使用的参数。

4. 查看 ACL

```
Router#show access-lists
```

5. 查看接口的配置

```
Router#show ip interface
```

6. 清除 ACL 的统计信息

```
Router#clear access-list counters [access-list-number|name]
```

该命令可以单独使用，也可以与特定的 ACL 的编号或者名称一起使用。

12.3 方案设计

在图 12-1 所示的网络拓扑中，因为 telnet 用的 TCP 端口号是 23，要实现 192.168.1.0/24 网络中的设备无法远程登录路由器 R2，可以采用扩展 ACL，根据源地址、目的地址、协议类型和端口号等进行数据包的访问控制。因为扩展 ACL 的放置位置要尽量靠近数据流量的起始点，所以需要在路由器 R1 上配置扩展 ACL。

12.4 项目实施

12.4.1 扩展编号 ACL 的配置

本小节网络拓扑如图 12-1 所示，路由器 R1 和 R2 的接口，以及计算机 PC1 和 PC2 的 IP 地址已经配置完成，要求完成扩展编号 ACL 的配置，实现 192.168.1.0/24 网络中的设备无法远程登录路由器 R2。

步骤 1：在路由器 R2 的全局配置模式下配置静态路由，输入以下代码。

```
R2(config)#ip route 192.168.0.0 255.255.255.0 10.10.0.1
R2(config)#ip route 192.168.1.0 255.255.255.0 10.10.0.1
```

步骤 2：在路由器 R2 的全局配置模式下配置远程登录，输入以下代码。

```
R2(config)#line vty 0 4
R2(config-line)#password 123
R2(config-line)#login
R2(config-line)#enable secret 123
```

步骤 3：进入计算机的"控制面板"窗口，如图 12-2 所示。

图 12-2 "控制面板"窗口

步骤 4：在"控制面板"窗口中单击"程序"选项，弹出"程序"窗口，如图 12-3 所示。

图 12-3 "程序"窗口

步骤 5：在"程序"窗口的"程序和功能"选项下单击"启用或关闭 Windows 功能"选项，弹出"Windows 功能"对话框，如图 12-4 所示。

图 12-4 "Windows 功能"对话框

步骤 6：在"Windows 功能"对话框的"启用或关闭 Windows 功能"界面中，选择"Telnet 客户端"复选项，然后单击"确定"按钮，提示"Windows 已完成请求的更改。"，如图 12-5 所示，启用 Telnet 客户端完成。

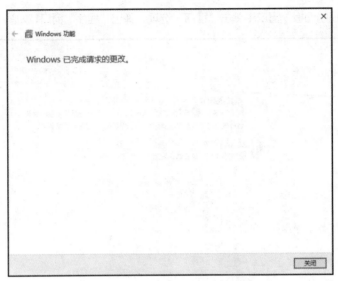

图 12-5 "Windows 已完成请求的更改。"提示

步骤 7：在计算机 PC2 的命令行界面输入 telnet 10.10.0.2，结果如图 12-6 所示。

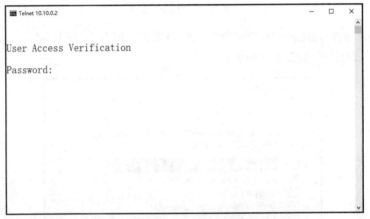

图 12-6 从计算机 PC2 远程登录路由器 R2 的结果

步骤 8：在路由器 R1 的全局配置模式下配置 ACL，输入以下代码。

```
R1(config)#access-list 100 deny tcp 192.168.1.0 0.0.0.255 host 10.10.0.2 eq 23
R1(config)#access-list 100 permit ip any any
```

步骤 9：在路由器 R1 的全局配置模式下在接口上应用 ACL，输入以下代码。

```
R1(config)#interface f0/1
R1(config-if)#ip access-group 100 in
```

步骤 10：在计算机 PC2 的命令行界面输入 telnet 10.10.0.2，如图 12-7 所示。

```
C:\>telnet 10.10.0.2
正在连接10.10.0.2...无法打开到主机的连接。 在端口 23: 连接失败
```

图 12-7 从计算机 PC2 远程登录路由器 R2

本小节在路由器 R2 上完成了远程登录的配置，通过查看从计算机 PC2 远程登录路由器 R2 的状态，我们发现，在配置 ACL 前，计算机 PC2 可以成功远程登录路由器 R2，完成 ACL 的配置后，计算机 PC2 无法成功远程登录路由器 R2。

12.4.2 扩展命名 ACL 的配置

本小节网络拓扑如图 12-1 所示，路由器 R1 和 R2 的接口和路由，以及计算机 PC1 和 PC2 的 IP 地址已经配置完成，已经实现了全网互通，要求完成扩展命名 ACL 的配置，实现 192.168.1.0/24 网络中的设备无法远程登录路由器 R2。

步骤 1：在路由器 R2 的全局配置模式下配置远程登录，输入以下代码。

```
R2(config)#line vty 0 4
R2(config-line)#password 123
R2(config-line)#login
R2(config-line)#enable secret 123
```

步骤 2：在计算机 PC2 的命令行界面输入 telnet 10.10.0.2，结果如图 12-8 所示。

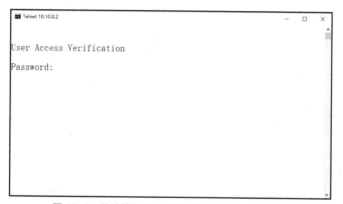

图 12-8　从计算机 PC2 远程登录路由器 R2 的结果

步骤 3：在路由器 R1 的全局配置模式下配置 ACL，输入以下代码。

```
R1(config)#ip access-list extended denytelnet
R1(config-ext-nacl)#deny tcp 192.168.1.0 0.0.0.255 host 10.10.0.2 eq 23
R1(config-ext-nacl)#permit ip any any
```

步骤 4：在路由器 R1 的全局配置模式下，在接口上应用 ACL，输入以下代码。

```
R1(config)# interface f0/1
R1(config-if)#ip access-group 100 in
```

步骤 5：在计算机 PC2 的命令行界面输入 telnet 10.10.0.2，如图 12-9 所示。

```
C:\>telnet 10.10.0.2
正在连接10.10.0.2...无法打开到主机的连接。 在端口 23: 连接失败
```

图 12-9　从计算机 PC2 远程登录路由器 R2

扩展 ACL 包括扩展编号 ACL 和扩展命名 ACL，本小节完成的是扩展命名 ACL 的配置。扩展命名 ACL 与扩展编号 ACL 的配置在命令格式方面稍微有点差别，两种形式的扩展 ACL 都可以实现相同的功能。

12.4.3 基于时间的 ACL 的配置

本小节网络拓扑如图 12-10 所示，计算机 PC1 位于学生网络，计算机 PC2 位于办公网络，路由器 R2 和计算机 PC3 位于学校外网。路由器 R1 和 R2 的接口，以及计算机 PC1、PC2 和 PC3 的 IP 地址已经配置完成，要求完成基于时间的 ACL 的配置，实现学生网络中的计算机在凌晨 0 点到早晨 6 点之间无法访问网络资源。

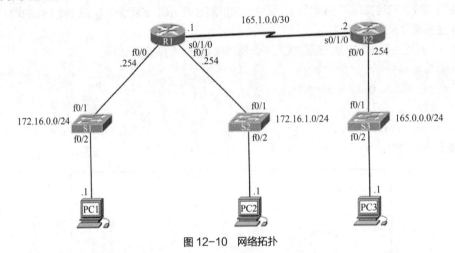

图 12-10 网络拓扑

步骤 1：在路由器 R1 的全局配置模式下配置 RIPv2，输入以下代码。

```
R1(config)#router rip
R1(config-router)#version 2
R1(config-router)#network 172.16.0.0
R1(config-router)#network 165.1.0.0
R1(config-router)#no auto-summary
```

步骤 2：在路由器 R2 的全局配置模式下配置 RIPv2，输入以下代码。

```
R2(config)#router rip
R2(config-router)#version 2
R2(config-router)#network165.0.0.0
R2(config-router)#network 165.1.0.0
R2(config-router)#no auto-summary
```

步骤 3：在计算机 PC1 的命令行界面输入 ping 172.16.1.1，检验联通性，如图 12-11 所示。

```
C:\>ping 172.16.1.1

正在 Ping 172.16.1.1 具有 32 字节的数据：
来自 172.16.1.1 的回复: 字节=32 时间<1ms TTL=127
来自 172.16.1.1 的回复: 字节=32 时间<1ms TTL=127
来自 172.16.1.1 的回复: 字节=32 时间<1ms TTL=127
来自 172.16.1.1 的回复: 字节=32 时间<1ms TTL=127

172.16.1.1 的 Ping 统计信息：
    数据包：已发送 = 4，已接收 = 4，丢失 = 0 (0% 丢失)，
往返行程的估计时间(以毫秒为单位)：
    最短 = 0ms，最长 = 0ms，平均 = 0ms
```

图 12-11 从计算机 PC1 ping 计算机 PC2

步骤 4：在计算机 PC1 的命令行界面输入 ping 165.0.0.1，检验联通性，如图 12-12 所示。

```
C:\>ping 165.0.0.1

正在 Ping 165.0.0.1 具有 32 字节的数据：
来自 165.0.0.1 的回复：字节=32 时间=10ms TTL=126
来自 165.0.0.1 的回复：字节=32 时间=10ms TTL=126
来自 165.0.0.1 的回复：字节=32 时间=10ms TTL=126
来自 165.0.0.1 的回复：字节=32 时间=10ms TTL=126

165.0.0.1 的 Ping 统计信息：
    数据包：已发送 = 4，已接收 = 4，丢失 = 0 (0% 丢失)，
往返行程的估计时间(以毫秒为单位)：
    最短 = 10ms，最长 = 10ms，平均 = 10ms
```

图 12-12　从计算机 PC1 ping 计算机 PC3

步骤 5：在路由器 R1 的全局配置模式下定义 ACL 起作用的时间范围，输入以下代码。

```
R1(config)#time-range denystudent
R1(config-time-range)#periodic daily 0:0 to 6:0
```

步骤 6：在路由器 R1 的全局配置模式下配置 ACL，输入以下代码。

```
R1(config)# access-list 100 deny ip 172.16.0.0 0.0.0.255 any time-range denystudent
R1(config-time-range)#access-list 100 permit ip any any
```

步骤 7：在路由器 R1 的全局配置模式下在接口上应用 ACL，输入以下代码。

```
R1(config)# interface f0/0
R1(config-if)#ip access-group 100 in
```

步骤 8：在路由器 R1 的特权执行模式下输入 show clock，查看系统当前时间，如图 12-13 所示。

```
R1#show clock
*07:31:34.579 UTC Tue Jan 29 2019
```

图 12-13　查看系统当前时间

步骤 9：在计算机 PC1 的命令行界面输入 ping 172.16.1.1，检验联通性，如图 12-14 所示。

```
C:\>ping 172.16.1.1

正在 Ping 172.16.1.1 具有 32 字节的数据：
来自 172.16.1.1 的回复：字节=32 时间<1ms TTL=127
来自 172.16.1.1 的回复：字节=32 时间<1ms TTL=127
来自 172.16.1.1 的回复：字节=32 时间<1ms TTL=127
来自 172.16.1.1 的回复：字节=32 时间<1ms TTL=127

172.16.1.1 的 Ping 统计信息：
    数据包：已发送 = 4，已接收 = 4，丢失 = 0 (0% 丢失)，
往返行程的估计时间(以毫秒为单位)：
    最短 = 0ms，最长 = 0ms，平均 = 0ms
```

图 12-14　从计算机 PC1 ping 计算机 PC2

步骤 10：在计算机 PC1 的命令行界面输入 ping 165.0.0.1，检验联通性，如图 12-15 所示。

```
C:\>ping 165.0.0.1

正在 Ping 165.0.0.1 具有 32 字节的数据:
来自 165.0.0.1 的回复: 字节=32 时间=10ms TTL=126
来自 165.0.0.1 的回复: 字节=32 时间=10ms TTL=126
来自 165.0.0.1 的回复: 字节=32 时间=10ms TTL=126
来自 165.0.0.1 的回复: 字节=32 时间=10ms TTL=126

165.0.0.1 的 Ping 统计信息:
    数据包: 已发送 = 4,已接收 = 4,丢失 = 0 (0% 丢失),
往返行程的估计时间(以毫秒为单位):
    最短 = 10ms,最长 = 10ms,平均 = 10ms
```

图 12-15 从计算机 PC1 ping 计算机 PC3

步骤 11:在路由器 R1 的特权执行模式下修改系统时间,输入以下代码。

```
R1#clock set 3:0:0 29 Jan 2019
```

步骤 12:在计算机 PC1 的命令行界面输入 ping 172.16.1.1,检验联通性,如图 12-16 所示。

```
C:\>ping 172.16.1.1

正在 Ping 172.16.1.1 具有 32 字节的数据:
来自 172.16.0.254 的回复: 无法访问目标网。
来自 172.16.0.254 的回复: 无法访问目标网。
来自 172.16.0.254 的回复: 无法访问目标网。
来自 172.16.0.254 的回复: 无法访问目标网。

172.16.1.1 的 Ping 统计信息:
    数据包: 已发送 = 4,已接收 = 4,丢失 = 0 (0% 丢失),
```

图 12-16 从计算机 PC1 ping 计算机 PC2

步骤 13:在计算机 PC1 的命令行界面输入 ping 165.0.0.1,检验联通性,如图 12-17 所示。

```
C:\>ping 165.0.0.1

正在 Ping 165.0.0.1 具有 32 字节的数据:
来自 172.16.0.254 的回复: 无法访问目标网。
来自 172.16.0.254 的回复: 无法访问目标网。
来自 172.16.0.254 的回复: 无法访问目标网。
来自 172.16.0.254 的回复: 无法访问目标网。

165.0.0.1 的 Ping 统计信息:
    数据包: 已发送 = 4,已接收 = 4,丢失 = 0 (0% 丢失),
```

图 12-17 从计算机 PC1 ping 计算机 PC3

本小节通过配置扩展 ACL 的作用时间范围,控制 ACL 起作用的时间。通过检验联通性可以发现,当系统时间不在 ACL 起作用的时间范围内时,学生网络中的计算机可以访问办公网络和学校外网;当系统时间在 ACL 起作用的时间范围内时,学生网络中的计算机无法访问办公网络和学校外网。

12.5 项目小结

本项目完成了扩展 ACL 的配置和基于时间的 ACL 的配置,扩展 ACL 可以根据数据包的多种属性过滤 IP 数据包，可以更加精确地控制流量。扩展 ACL 有扩展命名 ACL 和扩展编号 ACL 两种。在配置 ACL 时至少要有一条 permit 语句，要注意 ACL 语句的顺序，语句内容相同、顺序不同的 ACL 语句的访问控制结果可能不同。为了使不需要的流量在流经网络之前被过滤掉，扩展 ACL 应尽可能靠近控制流量的起始点。配置基于时间的 ACL 时先定义 ACL 起作用的时间范围，然后在 ACL 中调用该时间范围。

12.6 拓展训练

本项目的拓展训练网络拓扑如图 12-18 所示，要求完成如下配置：

（1）完成计算机 IP 地址、路由器接口和路由的配置，实现计算机 PC1、PC2 和路由器 R2 互通；

（2）完成基于时间的 ACL 的配置，实现计算机 PC1 在每天上午 8 点到 12 点之间和下午 1 点半到 5 点之间无法浏览网页。

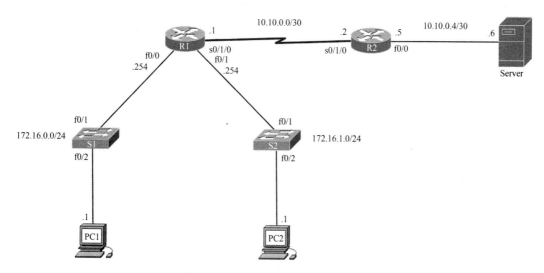

图 12-18　项目 12 拓展训练网络拓扑

模块四

网络地址转换的配置

计算机网络的主流 IP 地址是 IPv4 地址，但是，目前 IPv4 地址面临枯竭问题。网络地址转换被广泛应用于各种类型的网络中，它不仅解决了 IPv4 地址枯竭的问题，而且可以实现对外隐藏并保护网络内部的计算机的功能，从而有效地避免来自网络外部的攻击。本模块主要介绍静态 NAT 的配置、动态 NAT 的配置和基于端口的 NAT 的配置。

项目 13
静态NAT的配置

13.1 用户需求

某学校网络拓扑如图 13-1 所示,计算机 PC1 位于学校内网,路由器 R2 和计算机 PC2 位于学校外网,怎样使内网中配置了私有 IP 地址的设备访问公有网络(Internet)?要求内网的设备能够从学校外网被访问。

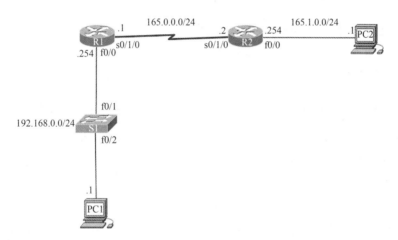

图 13-1 某学校网络拓扑

13.2 知识梳理

13.2.1 公有地址和私有地址

1. 公有地址

公有地址是必须在所属地域的相应 Internet 注册管理机构注册的地址,各种组织可以通过互联网服务提供商(Internet Service Provider,ISP)租用公有地址。公有地址可以直接通过 Internet 路由。

2. 私有地址

私有地址是保留的地址,是用于内网通信的一类 IP 地址,任何人均可以使用,

公有地址和私有地址

两个网络甚至更多的网络也可以使用相同的私有地址。为防止地址冲突，私有地址不能直接通过 Internet 路由。私有地址的分类及范围如表 13-1 所示。

表 13-1 私有地址的分类及范围

私有地址分类	私有地址范围
A 类	10.0.0.0～10.255.255.255
B 类	172.16.0.0～172.31.255.255
C 类	192.168.0.0～192.168.255.255

13.2.2 网络地址转换

网络地址转换（Network Address Translation，NAT）是一种 IP 地址转换技术，可以将不能直接在 Internet 上路由的私有地址转换成可路由的公有地址，让网络可以使用私有 IP 地址以节省公有 IP 地址。按照实现方式，NAT 可以分为静态 NAT、动态 NAT 和基于端口的 NAT（NAT 的过载）3 种。

网络地址转换

13.2.3 静态 NAT

静态 NAT 是指将内部网络的私有 IP 地址转换为公有 IP 地址，IP 地址采用一对一的转换，某个私有 IP 地址只能转换为某个公有 IP 地址。借助静态 NAT，可以实现外部网络对内部网络中某些特定设备（如服务器）的访问。

13.2.4 NAT 的术语

在图 13-2 所示的网络拓扑中，计算机 PC1 向服务器（Server）发送数据时启用了 NAT，把 192.168.1.1 转换成 165.0.0.3。根据地址在内部网络上还是在 Internet 上，以及流量是传入还是送出，不同的 IP 地址有不同的名称。

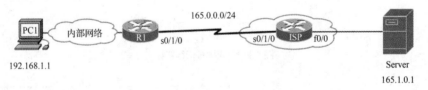

NAT 表			
内部本地地址	内部全局地址	外部本地地址	外部全局地址
192.168.1.1	165.0.0.3	165.1.0.1	165.1.0.1

图 13-2 NAT 网络拓扑

1. 内部本地地址

内部本地地址通常不是服务提供商分配的 IP 地址，一般是转换前本地主机的私有地址。图 13-2 中，IP 地址 192.168.1.1 被分配给内部网络上的主机 PC1，192.168.1.1 即是内部本地地址。

2. 内部全局地址

内部全局地址是当内部主机流量流出 NAT 路由器时被分配给内部主机的有效公有地址，一般是内部本地地址转换后的地址。图 13-2 中，当来自计算机 PC1 的流量发往服务器 165.1.0.1 时，首先由路由器 R1 进行地址转换，192.168.1.1 被转换为 165.0.0.3，165.0.0.3 即是内部全局地址。

3. 外部本地地址

外部本地地址是分配给外部网络上主机的本地 IP 地址。图 13-2 中，在外部网络中服务器本地的 IP 地址为 165.1.0.1，即外部本地地址。

4. 外部全局地址

外部全局地址是被分配给 Internet 上主机的可达 IP 地址。图 13-2 中，服务器的可达 IP 地址为 165.1.0.1。大多数情况下，外部本地地址与设备的外部全局地址相同。在有些网络中，为了隐藏公网中某些设备的真实 IP 地址，Internet 上主机可达的 IP 地址不是主机在本地网络中的 IP 地址，而是本地 IP 地址经过转换后的地址，通过访问转换后的这个地址可以访问 Internet 上的主机，这个地址就是外部全局地址。这种情况下，外部本地地址与设备的外部全局地址不同。

13.2.5 端口转发

端口转发是将数据包从一个网络节点转发至另一个网络节点的过程，可以将发往公有 IP 地址和路由器端口的数据包转发至私有 IP 地址和内部网络端口。端口转发实质上是指在特定 TCP 或 UDP 端口号上执行静态 NAT 的一种转换方式。

13.2.6 配置命令

1. 配置静态 NAT 的转换条目

```
Router(config)#ip nat inside source static local-ip global-ip
```

local-ip：内部本地地址。

global-ip：内部全局地址。

2. 指定内部接口

```
Router(config)#interface type number
Router(config-if)#ip nat inside
```

3. 指定外部接口

```
Router(config)#interface type number
Router(config-if)#ip nat outside
```

type number：接口的类型和编号。

4. 端口转发的配置

```
Router(config)#ip nat inside source static {tcp|udp} local-ip local-port global-ip global-port [extendable]
```

tcp│udp：指示该源是 TCP 或 UDP 端口号。

local-ip：分配给内部网络上的主机的 IPv4 地址。

local-port：TCP 或 UDP 端口号，是服务器侦听网络流量所使用的端口号，取值范围为 1～65535。

global-ip：内部主机的全局唯一 IPv4 地址，是外部客户端到达内部服务器所用的 IP 地址。

global-port：TCP 或 UDP 端口号，是外部客户端到达内部服务器所用的端口号，取值范围为 1～65535。

extendable：允许用户配置多个模糊的静态 NAT。模糊的静态 NAT 是指使用相同本地地址或全局地址的转换，它允许路由器必要时可在多个端口扩展转换。

5. 查看 NAT 表

```
Router#show ip nat translations
```

6. 清除 NAT 表

```
Router#clear ip nat translations
```

7. 查看 NAT 的统计信息

```
Router#show ip nat statistics
```

8. 清除 NAT 的统计信息

```
Router#clear ip nat statistics
```

13.3 方案设计

在图 13-1 所示的网络拓扑中，要实现配置了私有 IP 地址的设备能够访问 Internet，可以利用 NAT 把内网配置的私有 IP 地址在网络出口处转换为公有地址，使内网中配置了私有地址的计算机能够访问 Internet。因为静态 NAT 是一对一的转换，所以经过静态 NAT 后，从外网访问内部全局地址就可以访问内网中对应的设备。

13.4 项目实施

13.4.1 静态 NAT 的配置

本小节网络拓扑如图 13-1 所示，计算机 PC1 位于学校内网，路由器 R1 是内网的出口路由器，路由器 R2 和计算机 PC2 位于学校外网。路由器 R1 和 R2 的接口，以及计算机 PC1 和 PC2 的 IP 地址已经配置完成，要求完成静态 NAT 的配置，把 192.168.0.1 转换成 165.0.0.3。

步骤 1：在路由器 R1 的全局配置模式下配置默认路由，输入以下代码。

```
R1(config)#ip route 0.0.0.0 0.0.0.0 165.0.0.2
```

步骤 2：在路由器 R1 的全局配置模式下配置静态 NAT，输入以下代码。

```
R1(config)#ip nat inside source static 192.168.0.1 165.0.0.3
```

步骤 3：在路由器 R1 的全局配置模式下指定内部接口和外部接口，输入以下代码。

```
R1(config)#interface f0/0
R1(config-if)#ip nat inside
R1(config-if)#interface s0/1/0
R1(config-if)#ip nat outside
```

步骤 4：在计算机 PC1 的命令行界面输入 ping 165.1.0.1，检验联通性，如图 13-3 所示。

```
C:\>ping 165.1.0.1

正在 Ping 165.1.0.1 具有 32 字节的数据:
来自 165.1.0.1 的回复: 字节=32 时间=10ms TTL=126
来自 165.1.0.1 的回复: 字节=32 时间=10ms TTL=126
来自 165.1.0.1 的回复: 字节=32 时间=10ms TTL=126
来自 165.1.0.1 的回复: 字节=32 时间=10ms TTL=126

165.1.0.1 的 Ping 统计信息:
    数据包: 已发送 = 4，已接收 = 4，丢失 = 0 (0% 丢失)，
往返行程的估计时间(以毫秒为单位):
    最短 = 10ms，最长 = 10ms，平均 = 10ms
```

图 13-3 从计算机 PC1 ping 计算机 PC2

步骤 5：在路由器 R1 的特权执行模式下输入 show ip nat translations，查看 NAT 表，如图 13-4 所示。

```
R1#show ip nat translations
Pro Inside global      Inside local       Outside local      Outside global
icmp 165.0.0.3:56621   192.168.0.1:56621  165.1.0.1:56621    165.1.0.1:56621
icmp 165.0.0.3:57133   192.168.0.1:57133  165.1.0.1:57133    165.1.0.1:57133
icmp 165.0.0.3:57645   192.168.0.1:57645  165.1.0.1:57645    165.1.0.1:57645
icmp 165.0.0.3:57901   192.168.0.1:57901  165.1.0.1:57901    165.1.0.1:57901
--- 165.0.0.3          192.168.0.1        ---                ---
```

图 13-4　NAT 表

通过查看 NAT 表和检验联通性可以发现，内部本地地址 192.168.0.1 被转换成了内部全局地址 165.0.0.3，计算机 PC1 和 PC2 实现了互通。

13.4.2　端口转发的配置

本小节网络拓扑如图 13-5 所示，路由器 R1 位于内部网络，路由器 R2 是内部网络和外部网络的边界路由器，路由器 R3 位于外部网络。路由器 R1、R2 和 R3 的接口，以及路由器 R1 的 telnet 服务已经配置完成，要求完成端口转发的配置，把 192.168.0.1 的 23 端口映射为 165.0.0.1 的 2301 端口。

图 13-5　网络拓扑

步骤 1：在路由器 R1 的全局配置模式下配置静态路由，输入以下代码。

`R1(config)#ip route 165.0.0.0 255.255.255.0 192.168.0.2`

步骤 2：在路由器 R3 的全局配置模式下配置静态路由，输入以下代码。

`R3(config)#ip route 192.168.0.0 255.255.255.0 165.0.0.1`

步骤 3：在路由器 R2 的全局配置模式下配置静态 NAT 的转换关系，输入以下代码。

`R2(config)#ip nat inside source static tcp 192.168.0.1 23 165.0.0.1 2301`

步骤 4：在路由器 R2 的全局配置模式下指定内部接口和外部接口，输入以下代码。

`R2(config)#interface s0/1/0`

`R2(config-if)#ip nat inside`

`R2(config-if)#interface s0/1/1`

`R2(config-if)#ip nat outside`

步骤 5：在路由器 R3 的特权执行模式下输入 telnet 165.0.0.1 2301，检验配置，如图 13-6 所示。

```
R3#telnet 165.0.0.1 2301
Trying 165.0.0.1, 2301 ... Open

User Access Verification

Password:
```

图 13-6　检验配置

步骤 6：在路由器 R2 的特权执行模式下输入 show ip nat translations，查看 NAT 表，如图 13-7

所示。

```
R2#show ip nat translations
Pro Inside global      Inside local       Outside local       Outside global
tcp 165.0.0.1:2301     192.168.0.1:23     165.0.0.2:52141     165.0.0.2:52141
tcp 165.0.0.1:2301     192.168.0.1:23     ---                 ---
```

图 13-7 NAT 表

通过查看 NAT 表和从路由器 R3 远程登录路由器 R1 的状态可以发现，把 192.168.0.1 的 23 端口映射为 165.0.0.1 的 2301 端口后，从路由器 R3 远程登录 165.0.0.1 的 2301 端口，可以实现对路由器 R1 的远程登录。

13.5 项目小结

本项目完成了静态 NAT 和端口转发的配置，静态 NAT 实现的是内部本地地址和内部全局地址的一对一的转换，如果需要从外网访问内网中配置了私有 IP 地址的设备，则可以配置静态 NAT。

13.6 拓展训练

本项目的拓展训练网络拓扑如图 13-8 所示，服务器 Server 位于内网，路由器 R1 是内网的出口路由器，路由器 R2 和计算机 PC1 位于外网，要求完成如下配置：
（1）完成服务器 IP 地址、路由器接口和必要路由的配置；
（2）完成静态 NAT 的配置，要求从外网访问 209.165.0.3 时能够访问到服务器 Server。

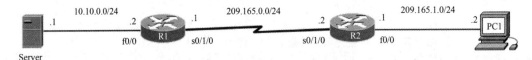

图 13-8 项目 13 拓展训练网络拓扑

项目 14
动态NAT的配置

14.1 用户需求

某学校网络拓扑如图 14-1 所示，计算机 PC1 位于学校内网，路由器 R2 和计算机 PC2 位于学校外网，内网中有 200 台计算机需要访问 Internet，目前已申请到 150 个公有 IP 地址，怎样使内网中配置了私有 IP 地址的所有设备访问 Internet？

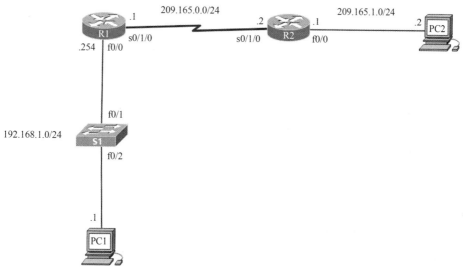

图 14-1 某学校网络拓扑

14.2 知识梳理

14.2.1 动态 NAT

动态 NAT

动态 NAT 是指将内部网络的一组私有 IP 地址转换为一组公有 IP 地址，转换时私有 IP 地址可随机转换为任何指定的合法 IP 地址。动态 NAT 用于在私有 IP 地址和公有 IP 地址间建立临时的、非固定的一对一或一对多的映射关系。当 ISP 提供的公有 IP 地址数量略少于网络内部的计算机数量时，可以采用动态 NAT。

14.2.2 配置命令

1. 定义公有地址池

```
Router(config)#ip nat pool name start-ip end-ip {netmask netmask|prefix-length prefix-length}
```

name：公有地址池的名称。
start-ip：公有地址池的起始地址。
end-ip：公有地址池的结束地址。
netmask：子网掩码。
prefix-length：前缀长度。

2. 定义标准 ACL

```
Router(config)#access-list access-list-number permit source [source-wildcard]
```

3. 建立动态源地址转换

```
Router(config)#ip nat inside source list access-list-number pool name
```

access-list-number：ACL 的编号，必须与定义的 ACL 编号一致。
name：公有地址池的名称，必须与定义的公有地址池的名称一致。

4. 指定内部接口

```
Router(config)#interface type number
Router(config-if)#ip nat inside
```

5. 指定外部接口

```
Router(config)#interface type number
Router(config-if)#ip nat outside
```

type number：接口的类型和编号。

14.3 方案设计

在图 14-1 所示的网络拓扑中，内网中需要访问 Internet 的设备数量比公有 IP 地址的数量多，但是相差不大，这种情况下，要实现内网中配置了私有 IP 地址的设备访问 Internet，可以利用动态 NAT 把内网配置的私有 IP 地址在网络出口处转换为公有 IP 地址。

14.4 项目实施

本节网络拓扑如图 14-1 所示，计算机 PC1 位于学校内网，路由器 R1 是内网的出口路由器，路由器 R2 和计算机 PC2 位于学校外网，目前申请到的公有 IP 地址的范围是 209.165.0.10～209.165.0.159，计算机的 IP 地址和路由器的接口已经配置完成，要求完成动态 NAT 的配置，实现内网中配置了私有 IP 地址的设备可以访问 Internet。

步骤 1：在路由器 R1 的全局配置模式下配置默认路由，输入以下代码。

```
R1(config)#ip route 0.0.0.0 0.0.0.0 209.165.0.2
```

步骤 2：在路由器 R1 的全局配置模式下创建公有地址池，输入以下代码。

```
R1(config)#ip nat pool pool1 209.165.0.10 209.165.0.159 netmask
```

255.255.255.0

步骤 3：在路由器 R1 的全局配置模式下配置 ACL，输入以下代码。

```
R1(config)#access-list 1 permit 192.168.1.0 0.0.0.255
```

步骤 4：在路由器 R1 的全局配置模式下配置动态 NAT 的转换关系，输入以下代码。

```
R1(config)#ip nat inside source list 1 pool pool1
```

步骤 5：在路由器 R1 的全局配置模式下指定内部接口和外部接口，输入以下代码。

```
R1(config)#interface f0/0
R1(config-if)#ip nat inside
R1(config-if)#interface s0/1/0
R1(config-if)#ip nat outside
```

步骤 6：在计算机 PC1 的命令行界面输入 ping 209.165.1.2，检验联通性，如图 14-2 所示。

```
C:\>ping 209.165.1.2

正在 Ping 209.165.1.2 具有 32 字节的数据：
来自 209.165.1.2 的回复：字节=32 时间=10ms TTL=126
来自 209.165.1.2 的回复：字节=32 时间=10ms TTL=126
来自 209.165.1.2 的回复：字节=32 时间=10ms TTL=126
来自 209.165.1.2 的回复：字节=32 时间=10ms TTL=126

209.165.1.2 的 Ping 统计信息：
    数据包：已发送 = 4，已接收 = 4，丢失 = 0（0% 丢失），
往返行程的估计时间(以毫秒为单位)：
    最短 = 10ms，最长 = 10ms，平均 = 10ms
```

图 14-2　从计算机 PC1 ping 计算机 PC2

步骤 7：在路由器 R1 的特权执行模式下输入 show ip nat translations，查看 NAT 表，如图 14-3 所示。

```
R1#show ip nat translations
Pro Inside global      Inside local       Outside local      Outside global
icmp 209.165.0.10:62946 192.168.1.1:62946 209.165.1.2:62946 209.165.1.2:62946
icmp 209.165.0.10:63458 192.168.1.1:63458 209.165.1.2:63458 209.165.1.2:63458
icmp 209.165.0.10:63970 192.168.1.1:63970 209.165.1.2:63970 209.165.1.2:63970
icmp 209.165.0.10:64226 192.168.1.1:64226 209.165.1.2:64226 209.165.1.2:64226
--- 209.165.0.10       192.168.1.1        ---                ---
```

图 14-3　NAT 表

通过查看 NAT 表和检验联通性可以发现，内部本地地址 192.168.1.1（192.168.1.0/24 网络中的地址）被转换成了内部全局地址 209.165.0.10（在 209.165.0.10～209.165.0.159 范围内），计算机 PC1 和 PC2 实现了互通。

14.5　项目小结

本项目完成了动态 NAT 的配置。动态 NAT 实现的是内部本地地址和内部全局地址的多对多的转换，当内网中配置私有地址的计算机比较多，申请到的公有地址比需要的地址少，但是缺少的数量不是很多时，可以采用动态 NAT。

14.6 拓展训练

本项目的拓展训练网络拓扑如图 14-4 所示，计算机 PC1、PC2 和路由器 R1 位于内网，路由器 R2 是内网的出口路由器，路由器 R3 和计算机 PC3 位于外网，要求完成如下配置：

（1）完成计算机 IP 地址、路由器接口和必要路由的配置。

（2）完成动态 NAT 的配置，要求把 192.168.0.0/24 和 192.168.1.0/24 网络中的地址转换成 165.0.0.10～165.0.0.254 之间的地址。

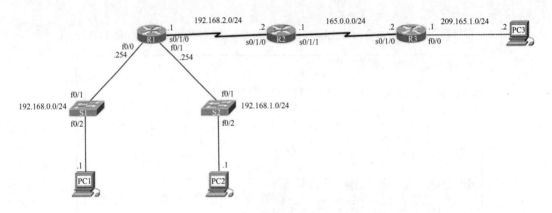

图 14-4 项目 14 拓展训练网络拓扑

项目 15
基于端口的NAT的配置

15.1 用户需求

某学校网络拓扑如图 15-1 所示，计算机 PC1、PC2 和路由器 R1 位于内网，路由器 R2 是内网的出口路由器，路由器 R3 和计算机 PC3 位于外网，目前只申请到一个公有 IP 地址，怎样使内网中配置了私有 IP 地址的设备访问 Internet？

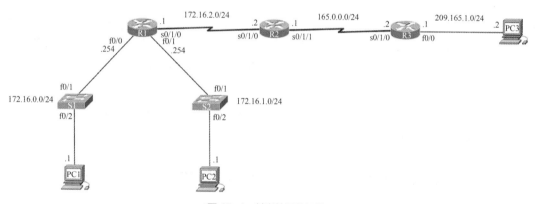

图 15-1 某学校网络拓扑

15.2 知识梳理

15.2.1 基于端口的 NAT

基于端口的 NAT 即端口地址转换（Port Address Translation，PAT），又称为 NAT 的过载，是指利用不同的端口号将多个内部 IP 地址转换为一个外部 IP 地址，从而可以最大限度地节约 IP 地址资源，同时可隐藏网络内部的所有主机，有效避免来自 Internet 的攻击。目前网络中应用最多的就是基于端口的 NAT。

基于端口的 NAT

15.2.2 配置命令

1. 定义标准 ACL

```
Router(config)#access-list access-list-number permit source [source-wildcard]
```

2. 建立动态源地址转换

```
Router(config)#ip nat inside source list access-list-number interface type number overload
```

access-list-number：ACL 的编号，必须与定义的 ACL 编号一致。

type number：接口的类型和编号，这个接口是连接到外网的接口。若仅有一个公有 IP 地址，那么在配置基于端口的 NAT 时，通常把该公有 IP 地址分配给连接到 ISP 的外部接口。所有内部地址离开该外部接口时，均被转换为该公有 IP 地址。

3. 指定内部接口

```
Router(config)#interface type number
Router(config-if)#ip nat inside
```

4. 指定外部接口

```
Router(config)#interface type number
Router(config-if)#ip nat outside
```

15.3 方案设计

在图 15-1 所示的网络拓扑中，只申请到一个公有 IP 地址，这种情况下要实现内网中配置了私有 IP 地址的设备访问 Internet，可以把这个公有 IP 地址配置到连接外网的接口，利用基于端口的 NAT 把内网配置的私有 IP 地址在网络出口处都转换成这个公有 IP 地址，这样可以最大限度地节约公有 IP 地址。

15.4 项目实施

15.4.1 基于端口的 NAT 的配置（单个内部接口）

本小节网络拓扑如图 15-1 所示，计算机 PC1、PC2 和路由器 R1 位于内网，路由器 R2 是内网的出口路由器，路由器 R3 和计算机 PC3 位于外网，目前申请到的 IP 地址是 165.0.0.1，路由器 R1、R2 和 R3 的接口，以及计算机 PC1、PC2 和 PC3 的 IP 地址已经配置完成，要求完成基于端口的 NAT 的配置，实现内网中配置了私有 IP 地址的设备可以访问 Internet。

步骤 1：在路由器 R1 的全局配置模式下配置默认路由，输入以下代码。

```
R1(config)#ip route 0.0.0.0 0.0.0.0 172.16.2.2
```

步骤 2：在路由器 R2 的全局配置模式下配置默认路由和下一跳静态路由，输入以下代码。

```
R2(config)#ip route 0.0.0.0 0.0.0.0 165.0.0.2
R2(config)#ip route 172.16.0.0 255.255.254.0 172.16.2.1
```

步骤 3：在路由器 R2 的全局配置模式下配置 ACL，输入以下代码。

```
R2(config)#access-list 1 permit 172.16.0.0 0.0.1.255
```

步骤 4：在路由器 R2 的全局配置模式下配置 NAT 的转换关系，输入以下代码。

```
R2(config)#ip nat inside source list 1 interface s0/1/1 overload
```

步骤 5：在路由器 R2 的全局配置模式下指定内部接口和外部接口，输入以下代码。

```
R2(config)#interface s0/1/0
```

```
R2(config-if)#ip nat inside
R2(config-if)#interface s0/1/1
R2(config-if)#ip nat outside
```

步骤 6：在计算机 PC1 的命令行界面输入 ping 209.165.1.2，检验联通性，如图 15-2 所示。

```
C:\>ping 209.165.1.2

正在 Ping 209.165.1.2 具有 32 字节的数据：
来自 209.165.1.2 的回复：字节=32 时间=19ms TTL=125
来自 209.165.1.2 的回复：字节=32 时间=19ms TTL=125
来自 209.165.1.2 的回复：字节=32 时间=19ms TTL=125
来自 209.165.1.2 的回复：字节=32 时间=19ms TTL=125

209.165.1.2 的 Ping 统计信息：
    数据包：已发送 = 4，已接收 = 4，丢失 = 0 (0% 丢失)，
往返行程的估计时间(以毫秒为单位)：
    最短 = 19ms，最长 = 19ms，平均 = 19ms
```

图 15-2　从计算机 PC1 ping 计算机 PC3

步骤 7：在计算机 PC2 的命令行界面输入 ping 209.165.1.2，检验联通性，如图 15-3 所示。

```
C:\>ping 209.165.1.2

正在 Ping 209.165.1.2 具有 32 字节的数据：
来自 209.165.1.2 的回复：字节=32 时间=18ms TTL=125
来自 209.165.1.2 的回复：字节=32 时间=18ms TTL=125
来自 209.165.1.2 的回复：字节=32 时间=18ms TTL=125
来自 209.165.1.2 的回复：字节=32 时间=18ms TTL=125

209.165.1.2 的 Ping 统计信息：
    数据包：已发送 = 4，已接收 = 4，丢失 = 0 (0% 丢失)，
往返行程的估计时间(以毫秒为单位)：
    最短 = 18ms，最长 = 18ms，平均 = 18ms
```

图 15-3　从计算机 PC2 ping 计算机 PC3

步骤 8：在路由器 R2 的特权执行模式下输入 show ip nat translations，查看 NAT 表，如图 15-4 所示。

```
R2#show ip nat translations
Pro Inside global       Inside local        Outside local       Outside global
icmp 165.0.0.1:48700    172.16.0.1:48700    209.165.1.2:48700   209.165.1.2:48700
icmp 165.0.0.1:48956    172.16.0.1:48956    209.165.1.2:48956   209.165.1.2:48956
icmp 165.0.0.1:49212    172.16.0.1:49212    209.165.1.2:49212   209.165.1.2:49212
icmp 165.0.0.1:49468    172.16.0.1:49468    209.165.1.2:49468   209.165.1.2:49468
icmp 165.0.0.1:45628    172.16.1.1:45628    209.165.1.2:45628   209.165.1.2:45628
icmp 165.0.0.1:45884    172.16.1.1:45884    209.165.1.2:45884   209.165.1.2:45884
icmp 165.0.0.1:46140    172.16.1.1:46140    209.165.1.2:46140   209.165.1.2:46140
icmp 165.0.0.1:46396    172.16.1.1:46396    209.165.1.2:46396   209.165.1.2:46396
```

图 15-4　NAT 表

通过查看 NAT 表和检验联通性可以发现，内部本地地址 172.16.0.1（172.16.0.0/24 网络中的地址）和 172.16.1.1（172.16.1.0/24 网络中的地址）都被转换成了内部全局地址 209.165.1.2，不同的转换，端口号不同，计算机 PC1 和 PC3、PC2 和 PC3 都实现了互通。

15.4.2　基于端口的 NAT 的配置（多个内部接口）

本小节网络拓扑如图 15-5 所示，计算机 PC1 和 PC2 位于内网，路由器 R1 是内网的出口路由器，

路由器 R2 和计算机 PC3 位于外网。目前申请到的 IP 地址是 165.0.0.1，路由器 R1 和 R2 的接口，以及计算机 PC1、PC2 和 PC3 的 IP 地址已经配置完成，要求完成基于端口的 NAT 的配置，实现内网中配置了私有 IP 地址的设备可以访问 Internet。

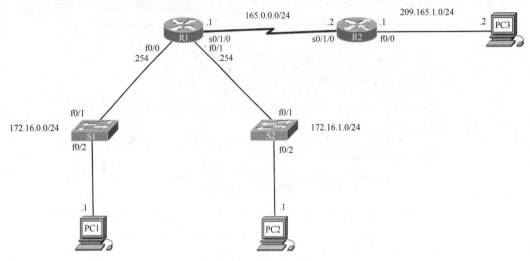

图 15-5　网络拓扑

步骤 1：在路由器 R1 的全局配置模式下配置默认路由，输入以下代码。

```
R1(config)#ip route 0.0.0.0 0.0.0.0 165.0.0.2
```

步骤 2：在路由器 R1 的全局配置模式下配置 ACL，输入以下代码。

```
R1(config)#access-list 1 permit 172.16.0.0 0.0.1.255
```

步骤 3：在路由器 R1 的全局配置模式下配置基于端口的 NAT，输入以下代码。

```
R1(config)#ip nat inside source list 1 interface s0/1/0 overload
```

步骤 4：在路由器 R1 的全局配置模式下指定内部接口和外部接口，输入以下代码。

```
R1(config)#interface f0/0
R1(config-if)#ip nat inside
R1(config-if)#interface f0/1
R1(config-if)#ip nat inside
R1(config-if)#interface s0/1/0
R1(config-if)#ip nat outside
```

步骤 5：在计算机 PC1 的命令行界面输入 ping 209.165.1.2，检验联通性，如图 15-6 所示。

```
C:\>ping 209.165.1.2
正在 Ping 209.165.1.2 具有 32 字节的数据:
来自 209.165.1.2 的回复: 字节=32 时间=10ms TTL=126
来自 209.165.1.2 的回复: 字节=32 时间=10ms TTL=126
来自 209.165.1.2 的回复: 字节=32 时间=10ms TTL=126
来自 209.165.1.2 的回复: 字节=32 时间=10ms TTL=126

209.165.1.2 的 Ping 统计信息:
    数据包: 已发送 = 4, 已接收 = 4, 丢失 = 0 (0% 丢失),
往返行程的估计时间(以毫秒为单位):
    最短 = 10ms, 最长 = 10ms, 平均 = 10ms
```

图 15-6　从计算机 PC1 ping 计算机 PC3

步骤 6：在计算机 PC2 的命令行界面输入 ping 209.165.1.2，检验联通性，如图 15-7 所示。

```
C:\>ping 209.165.1.2

正在 Ping 209.165.1.2 具有 32 字节的数据:
来自 209.165.1.2 的回复: 字节=32 时间=9ms TTL=126
来自 209.165.1.2 的回复: 字节=32 时间=9ms TTL=126
来自 209.165.1.2 的回复: 字节=32 时间=9ms TTL=126
来自 209.165.1.2 的回复: 字节=32 时间=9ms TTL=126

209.165.1.2 的 Ping 统计信息:
    数据包: 已发送 = 4，已接收 = 4，丢失 = 0 (0% 丢失)，
往返行程的估计时间(以毫秒为单位):
    最短 = 9ms，最长 = 9ms，平均 = 9ms
```

图 15-7　从计算机 PC2 ping 计算机 PC3

步骤 7：在路由器 R1 的特权执行模式下输入 show ip nat translations，查看 NAT 表，如图 15-8 所示。

```
R1#show ip nat translations
Pro Inside global      Inside local         Outside local        Outside global
icmp 165.0.0.1:46409   172.16.0.1:46409     209.165.1.2:46409    209.165.1.2:46409
icmp 165.0.0.1:46665   172.16.0.1:46665     209.165.1.2:46665    209.165.1.2:46665
icmp 165.0.0.1:46921   172.16.0.1:46921     209.165.1.2:46921    209.165.1.2:46921
icmp 165.0.0.1:47177   172.16.0.1:47177     209.165.1.2:47177    209.165.1.2:47177
icmp 165.0.0.1:51529   172.16.1.1:51529     209.165.1.2:51529    209.165.1.2:51529
icmp 165.0.0.1:51785   172.16.1.1:51785     209.165.1.2:51785    209.165.1.2:51785
icmp 165.0.0.1:52041   172.16.1.1:52041     209.165.1.2:52041    209.165.1.2:52041
icmp 165.0.0.1:52297   172.16.1.1:52297     209.165.1.2:52297    209.165.1.2:52297
```

图 15-8　NAT 表

> **注意**　如果出口路由器有多个接口互联了内网，并且这些内网的 IP 地址都需要进行 NAT，那么在配置 NAT 时，首先用标准 ACL 把需要转换的多个内网地址抓取出来，然后在每个连接内网的接口处都输入 ip nat inside 命令。

15.5　项目小结

本项目完成了基于端口的 NAT 的配置，基于端口的 NAT 利用不同的端口号将多个内部 IP 地址转换为一个外部 IP 地址。当内网中配置了私有地址的计算机比较多，公有地址只有一个时，可以采用基于端口的 NAT 把内网的私有地址都转换成这个公有地址，这样可以最大限度地节约公有地址。出口路由器如果有多个连接内网的接口，在配置 NAT 时，每个内部接口处都要输入 ip nat inside 命令。

15.6　拓展训练

本项目的拓展训练网络拓扑如图 15-9 所示，计算机 PC1、PC2 和服务器 Server 位于内网，路由器 R1 是内网的出口路由器，路由器 R2 和计算机 PC3 位于外网，要求完成如下配置：

（1）完成服务器 IP 地址、路由器接口和必要路由的配置；

(2)完成基于端口的 NAT 的配置,实现把 192.168.0.0/24 和 192.168.1.0/24 网络中的地址转换为路由器 R1 的 s0/1/0 接口的地址;

(3)完成静态 NAT 的配置,实现能够从外网通过 165.1.0.3 地址访问服务器 Server。

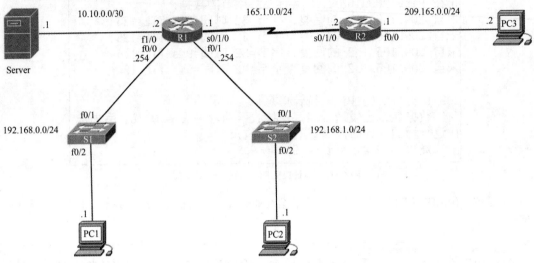

图 15-9　项目 15 拓展训练网络拓扑

模块五

网络优化与安全的配置

网络的高速、不中断、可靠运行是衡量网络传输质量的重要指标，很多企业网络都要求网络不中断，在保证网络可靠运行的前提下，要求保证网络的传输速率。网络的优化与安全对提高网络传输质量、保证网络可靠运行非常重要。本模块主要介绍生成树的配置、链路聚合的配置和交换机端口安全的配置。

项目 16

生成树的配置

16.1 用户需求

某学校网络拓扑如图 16-1 所示,该网络由 3 台交换机和 3 台计算机组成,请分析这种网络拓扑结构的优点和缺点,以及要保证网络正常运行需要做哪些配置。

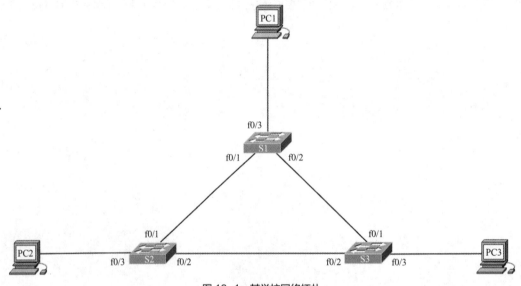

图 16-1 某学校网络拓扑

16.2 知识梳理

16.2.1 网络冗余

网络冗余和网络中的二层环路

如图 16-1 所示,如果网络中只有交换机 S1 和 S2,没有交换机 S3,计算机 PC1 和 PC2 之间只能通过 PC1-S1-S2-PC2 这条链路通信,只要这条链路上有一台设备或者一条线缆出现故障,计算机 PC1 和 PC2 的通信就无法完成。添加了交换机 S3 和相应的线缆后,计算机 PC1 与 PC2 既可以通过 PC1-S1-S2-PC2 这条链路通信,也可以通过 PC1-S1-S3-S2-PC2 这条链路通信,即使一条链路上的设备或者线缆出现故障,只要另外一条链路正常运行，依然可以保持计算机 PC1 与 PC2 互通。对于计算机 PC1 和 PC2 的通信，交换机 S3 及交换机 S3 与交换机 S1 和 S2 互联的链路是网络中的冗余设备和冗余线路，即网络冗余，可以提高通信的可靠性。

16.2.2 网络中的二层环路

网络冗余可以提供备份路径，可以确保网络的可用性，但网络冗余也会带来二层环路。如果将广播帧从除源端口之外的所有交换机端口转发出去，由于可转发该帧的路径不止一条，因此可能会导致网络广播帧的无尽循环。与通过路由器传递的 IP 数据报文不同，以太网中的帧不含 TTL 字段，如果交换网络中的帧没有正确终止，那么它们就会在交换机之间无休止地传输，直到链路断开或环路解除为止。

在图 16-1 所示的网络拓扑中，交换机 S1、S2 和 S3 之间存在环路。此时如果一个广播帧由某台交换机转发，就会出现该帧在 3 台交换机 S1、S2 和 S3 上循环转发的情况，这就产生了二层环路。环路中的所有交换机之间不断相互发送相同的帧，这使得交换机的 CPU 不得不处理大量的数据，会导致参与环路的所有交换机的 CPU 负载过高，使交换机无法高效处理收到的正常流量。由于 MAC 地址表不断被广播帧的内容更新，因此交换机不知道究竟使用哪个端口才能将单播帧转发到最终目的地，这造成单播帧也在网络中不断循环。在网络中循环的帧越来越多，便形成了广播风暴，导致被卷入环路的主机无法被网络中的其他主机访问。

16.2.3 生成树协议

生成树协议（Spanning Tree Protocol，STP）能够阻塞可能导致环路的冗余链路，以确保网络中的所有目的地之间只有一条逻辑路径。当端口处于阻塞状态时，流量将无法进入或流出，但 STP 用来防止环路的网桥协议数据单元（Bridge Protocol Data Unit，BPDU）帧仍可继续通行。为了提供冗余功能，这些物理路径实际上依然存在，只是被禁用。一旦需要启用此类路径来抵消网络线缆或交换机故障的影响，STP 就会重新计算路径，将必要的端口解除阻塞，使冗余路径进入活动状态。

生成树协议

16.2.4 STP 的类型

STP 的类型主要有 4 种，分别是 STP、PVST+、RSTP 和 MSTP。

1. STP

STP 是原始的 IEEE 802.1D 版本（IEEE 802.1D-1998 及更早版本），在具有冗余链路的网络中提供无环拓扑。对于这个版本的生成树协议，不论 VLAN 的数量如何，整个桥接网络中都只有一个生成树实例，因此消耗的 CPU 和内存资源比其他版本的少一些。

2. PVST+

PVST+（Per-VLAN Spanning Tree Plus）是思科对 STP 做了一些改进后创建的，它为网络中配置的每个 VLAN 都提供单独的 IEEE 802.1D 生成树实例，这个独立的实例支持很多增强特性，如 PortFast、UplinkFast、BackboneFast、BPDU 防护、BPDU 过滤、根防护和环路防护。

3. RSTP

快速生成树协议（Rapid Spanning Tree Protocol，RSTP）从 STP 演变而来，是 STP 的进化版本，它可以使 STP 收敛得更快，收敛速度快于 STP。

4. MSTP

多生成树协议（Multiple Spanning Tree Protocol，MSTP）是一个 IEEE 标准，可将多个 VLAN 映射到同一个生成树实例。思科采用的 MSTP 提供多达 16 个实例，并将许多具有相同物理和逻辑拓扑的 VLAN 合并到一个常用实例中，而且每个实例都支持 PortFast、BPDU 防护、BPDU 过滤、根防护和环路防护。

16.2.5　STP 的算法

STP 使用生成树算法使网络中的某些交换机端口处于阻塞状态，防止环路形成。生成树算法会将一台交换机指定为根桥，然后将其作为计算所有路径的参考点。确定根桥后，生成树算法会计算到根桥的最佳路径。在生成树算法为广播域中的所有目的地确定到达根桥的最佳路径时，网络中的所有流量都会停止转发。生成树算法在确定要开放的路径时，会同时考虑路径开销和端口开销。路径开销是根据端口开销计算出来的，而端口开销与给定路径上的每个交换机端口的端口速度相关联。端口开销的总和决定了到达根桥的路径总开销。如果可供选择的路径不止一条，那么生成树算法会选择路径总开销最小的路径。生成树算法确定了哪些路径保留为可用路径之后，会将交换机端口配置为不同的端口角色。端口角色描述了网络中的端口与根桥的关系，以及端口是否能转发流量。

16.2.6　STP 的端口角色

1. 根端口

根端口为最靠近根桥的交换机端口。根端口存在于非根桥上，该端口具有到根桥的最佳路径。根端口向根桥转发流量。根端口可以使用所接收帧的源 MAC 地址填充 MAC 地址表。一个网段只能有一个根端口。

2. 指定端口

指定端口为网络中被获准转发流量的、除根端口之外的所有端口。指定端口存在于根桥和非根桥上。根桥上的所有交换机端口都是指定端口。而对于非根桥，指定端口是指根据需要接收帧或向根桥转发帧的交换机端口，一个网段只能有一个指定端口。

3. 非指定端口

非指定端口是为防止环路而被阻塞的交换机端口，此类端口不会转发数据帧，也不使用源地址填充 MAC 地址表。在某些 STP 的衍生协议中，非指定端口又称为替换端口。

4. 禁用端口

禁用端口是指被关闭掉的交换机端口。

16.2.7　网桥 ID

网桥 ID（Bridge ID，BID）包含 3 个字段：网桥优先级、扩展系统 ID、MAC 地址。

1. 网桥优先级

网桥优先级的值可自定义，优先级最低的交换机具有最小的 BID 值，这样的交换机会成为根桥（网桥优先级值越小，优先级越高）。思科交换机的默认网桥优先级值是 32768。网桥优先级的取值范围是 1~65536，1 表示最高优先级。

2. 扩展系统 ID

扩展系统 ID 字段包含的是与 BPDU 关联的 VLAN ID，与网桥优先级一起可标识 BPDU 帧的优先级及其所属的 VLAN。在特定配置下，BPDU 帧可能不含扩展系统 ID。早期的 STP 用于不使用 VLAN 的网络中，所有交换机构成一棵简单的生成树，这时不包含扩展系统 ID。当 VLAN 逐渐成为常用的网络架构分段方式时，人们对 STP 进行了改进，加入了对 VLAN 的支持。使用扩展系统 ID 时，网桥优先级值的可用位数会随之改变，网桥优先级值的增量从 1 变为 4096，所以网桥优先级值只能是 4096 的倍数。

3. MAC 地址

当两台交换机配置了相同的网桥优先级和相同的扩展系统 ID 时，MAC 地址所含的十六进制数最小的交换机具有较小的 BID 值。当交换机具有相同的默认网桥优先级值时，MAC 地址就成为确定交换机能否成为根桥的决定因素。

16.2.8 STP 的收敛

1. 选举根桥

网络中对应的每个 VLAN 只能有一个网桥作为根桥，BID 值最小的网桥当选为根桥。根桥上的所有端口都将成为指定端口。

2. 选举根端口

STP 会在每个非根桥上选举一个根端口，根端口所连接的路径是非根桥到根桥之间开销最小的路径。当同一交换机上有两个以上的端口到根桥的路径开销相同时（有多条等价路径），非根桥会选择端口 ID 数值最小的端口作为根端口，端口 ID 由端口优先级和端口号组成，如果端口优先级相同，那么使用默认端口号来做出抉择，默认端口号最小的端口将成为根端口。

3. 选举指定端口和非指定端口

STP 会在网桥的每个网段选举一个指定端口，它是到达根桥路径开销最小的端口。当两个非根端口的交换机端口连接到同一个 LAN 网段时，非根桥到根桥的路径开销相同，会发生端口角色竞争，这两台交换机会交换 BPDU 帧，由 BID 决定谁成为指定端口。当交换机接收对方发来的 BID 时，会将其与自己本身的 BID 进行比较，如果接收的 BID 的值比自己本身的 BID 的值小，则对方的端口（发送方的端口）成为指定端口，反之交换机自己的端口成为指定端口；如果 BID 值也相同，则由端口 ID 决定谁成为指定端口。除了根端口和指定端口外的其他端口都将成为非指定端口，非指定端口最终会进入阻塞状态以防止生成环路。

当交换机检测到一个端口从阻塞状态变为转发状态或转发端口关闭（如被阻塞）时，交换机会认为自己检测到了拓扑更改，会通知生成树的根桥，然后根桥将该信息广播到整个网络。

16.2.9 路径成本

路径成本的计算和链路带宽相关联，根路径成本就是到根桥的路径中所有链路的路径成本的累计，不同的链路带宽对应的修订前后的 IEEE 802.1D 路径成本如表 16-1 所示。

表 16-1 路径成本

链路带宽	成本（修订前）	成本（修订后）
10 Gbit/s	1	2
1000 Mbit/s	1	4
100 Mbit/s	10	19
10 Mbit/s	100	100

16.2.10 端口状态

STP 用于为整个广播域确定逻辑无环路径。互联的交换机通过交换 BPDU 帧来获知信息，生成树是根据这些信息确定的。每个交换机的端口都会经过如下 5 种可能的状态。

1. 阻塞状态

处于阻塞（Blocking）状态的端口不参与数据的转发，不能把 MAC 地址加入 MAC 地址表。它可以接收 BPDU 帧，并将其交给 CPU 进行处理，不能发送 BPDU 帧。

2. 侦听状态

处于侦听（Listening）状态的交换机端口不能接收或者传输数据，也不能把 MAC 地址加入 MAC 地址表。它可以接收 BPDU 帧，也可以发送自己的 BPDU 帧。

3. 学习状态

处于学习（Learning）状态的交换机端口可以发送和接收 BPDU 帧，准备参与帧的转发。它虽然不能转发数据，但可以学习 MAC 地址，并开始填充 MAC 地址表。

4. 转发状态

处于转发（Forwarding）状态的端口是活动拓扑的一部分。该端口能够发送和接收数据，学习 MAC 地址，发送和接收 BPDU 帧。

5. 禁用状态

处于禁用（Disabled）状态的端口不参与生成树，不会转发帧。当管理性关闭交换机端口时，端口进入禁用状态。

16.2.11 RSTP 端口

RSTP 是 STP 的改进版，RSTP 能够在第 2 层网络拓扑变化时加速重新计算生成树的过程，RSTP 重新定义了端口角色及端口状态。它定义了另外两种端口角色，即替代端口和备份端口；定义了 3 种端口状态，即丢弃状态、学习状态和转发状态。如果端口被配置为替代端口或备份端口，则该端口可以立即进入转发状态，而无须等待网络收敛。

1. 端口角色

（1）替代端口

替代端口是用来提供去往根桥的替代路径的端口，替代端口在稳定工作状态的拓扑中处于丢弃状态。替代端口出现在非指定交换机上，并且会在当前的指定端口出现故障时变为指定端口。

（2）备份端口

备份端口是指定交换机上的一个额外的端口。它的作用是在当前指定端口失效时为其提供备份链路。它的端口 ID 数值大于指定交换机指定端口的端口 ID 数值，在稳定工作状态的拓扑中，备份端口处于丢弃状态。

2. 端口状态

（1）丢弃状态

稳定的活动拓扑及拓扑同步和更改期间都会出现此状态。该状态下禁止转发数据帧，因而可以断开二层环路。

（2）学习状态

处于学习状态的端口会接收数据帧来填充 MAC 地址表，以限制未知单播帧泛洪。稳定的活动拓扑及拓扑同步和更改期间都会出现此状态。

（3）转发状态

处于转发状态的交换机端口决定了拓扑，仅在稳定的活动拓扑中出现。

16.2.12 边缘端口

边缘端口是指永远不会用于连接其他交换机的端口，它直接连接终端。启用时，边缘端口会跳过耗时的侦听和学习等状态，立即进入转发状态。如果把边缘端口连接到其他交换机，则可能会发生环路，并可能会延迟生成树的收敛。

16.2.13 BPDU 防护

配置了 PortFast 的边缘端口不接收 BPDU 帧。如果端口收到了 BPDU 帧，就意味着另一个网桥或交换机已连接该端口，从而可能导致生成树环路。思科交换机支持 BPDU 防护功能。当端口启用了 BPDU 防护时，会在收到 BPDU 帧时将端口设置为 error-disabled 状态，有效关闭端口。端口一旦进入 error-disabled 状态，只能手动恢复。

> 📖 边学边思
> 　　我们要树立绿色发展的理念，以确保网络的健康和稳定运行。绿色发展强调在网络建设和运营中考虑运行环境的可持续性，并积极采取措施减少能源消耗等。

16.2.14 配置命令

1. 配置交换机的网桥优先级

```
Switch(config)#spanning-tree vlan vlan-id priority value
```

value：网桥优先级的值，介于 0 和 65536 之间，增量为 4096。

2. 将交换机配置为根桥

```
Switch(config)#spanning-tree vlan-id root primary
```

将交换机的网桥优先级值配置为预定义的值 24576，或者是比网络中检测到的最小网桥优先级值小 4096 的值。

3. 将交换机配置为备份根桥

```
Switch(config)#spanning-tree vlan-id root secondary
```

将交换机的网桥优先级值设置为预定义的值 28672，这可确保在主根桥失败的情况下，该交换机能在新一轮的根桥选举中成为根桥。

4. 配置端口优先级

```
Switch(config-if)#spanning-tree port-priority value
```

value：端口优先级的值，范围为 0~240（增量为 16）。默认的端口优先级值是 128。与网桥优先级一样，端口优先级值越小，优先级越高。

5. 配置端口的路径成本

```
Switch(config-if)#spanning-tree cost cost
```

6. 查看生成树的配置

```
Switch#show spanning-tree
Switch#show spanning-tree interface interface-id
```

7. 配置边缘端口

```
Switch(config)#interface interface-id
Switch(config-if)#switchport mode access
Switch(config-if)#spanning-tree portfast
```

8. 配置 BPDU 防护

```
Switch(config)#interface interface-id
Switch(config-if)#spanning-tree bpduguard enable
```

如果要在边缘端口上启用 BPDU 防护，则可以用如下命令：

```
Switch(config)# spanning-tree portfast bpduguard default
```

16.3 方案设计

在图 16-1 所示的网络拓扑中，交换机 S1、S2 和 S3 的物理链路连接成了环，这种有冗余的网络提高了网络的可靠性。与此同时，网络中存在二层环路，容易引起广播风暴。二层环路引起的广播风暴不会自动消除，最终将造成网络瘫痪。解决这种问题的方法是在交换机上启用 STP，在保证网络可靠性的同时通过计算阻塞特定的端口，从而消除网络中的二层环路。可以通过 STP 的配置，控制网络中的根桥和端口角色。

16.4 项目实施

16.4.1 根桥和端口角色的查看

本小节网络拓扑如图 16-1 所示，交换机 S1 的 MAC 地址是 18ef.6394.3680，交换机 S2 的 MAC 地址是 18ef.638b.0500，交换机 S3 的 MAC 地址是 18ef.6366.7b80，网桥优先级值是默认值，要求分析并查看网络中的根桥、根端口、指定端口和非指定端口。

步骤 1：在交换机 S1 的特权执行模式下输入 show spanning-tree，查看生成树的配置，如图 16-2 所示。

```
S1#show spanning-tree
VLAN0001
  Spanning tree enabled protocol ieee
  Root ID    Priority    32769
             Address     18ef.6366.7b80
             Cost        19
             Port        2 (FastEthernet0/2)
             Hello Time  2 sec  Max Age 20 sec  Forward Delay 15 sec

  Bridge ID  Priority    32769  (priority 32768 sys-id-ext 1)
             Address     18ef.6394.3680
             Hello Time  2 sec  Max Age 20 sec  Forward Delay 15 sec
             Aging Time  15

Interface        Role Sts Cost      Prio.Nbr Type
---------------- ---- --- --------- -------- --------
Fa0/1            Altn BLK 19        128.1    P2p
Fa0/2            Root FWD 19        128.2    P2p
Fa0/3            Desg FWD 19        128.3    P2p
```

图 16-2 交换机 S1 生成树的配置

步骤 2：在交换机 S2 的特权执行模式下输入 show spanning-tree，查看生成树的配置，如图 16-3 所示。

```
S2#show spanning-tree
VLAN0001
  Spanning tree enabled protocol ieee
  Root ID    Priority    32769
             Address     18ef.6366.7b80
             Cost        19
             Port        2 (FastEthernet0/2)
             Hello Time  2 sec  Max Age 20 sec  Forward Delay 15 sec

  Bridge ID  Priority    32769  (priority 32768 sys-id-ext 1)
             Address     18ef.638b.0500
             Hello Time  2 sec  Max Age 20 sec  Forward Delay 15 sec
             Aging Time 300

Interface        Role Sts Cost      Prio.Nbr Type
---------------- ---- --- --------- -------- --------------------
Fa0/1            Desg FWD 19        128.1    P2p
Fa0/2            Root FWD 19        128.2    P2p
Fa0/3            Desg FWD 19        128.3    P2p
```

图 16-3　交换机 S2 生成树的配置

步骤 3：在交换机 S3 的特权执行模式下输入 show spanning-tree，查看生成树的配置，如图 16-4 所示。

```
S3#show spanning-tree
VLAN0001
  Spanning tree enabled protocol ieee
  Root ID    Priority    32769
             Address     18ef.6366.7b80
             This bridge is the root
             Hello Time  2 sec  Max Age 20 sec  Forward Delay 15 sec

  Bridge ID  Priority    32769  (priority 32768 sys-id-ext 1)
             Address     18ef.6366.7b80
             Hello Time  2 sec  Max Age 20 sec  Forward Delay 15 sec
             Aging Time 300

Interface        Role Sts Cost      Prio.Nbr Type
---------------- ---- --- --------- -------- --------------------
Fa0/1            Desg FWD 19        128.1    P2p
Fa0/2            Desg FWD 19        128.2    P2p
Fa0/3            Desg FWD 19        128.3    P2p
```

图 16-4　交换机 S3 生成树的配置

结果分析如下。

（1）根桥的选举

根桥的选举依据 BID，普通 STP 的 BID 由网桥优先级和 MAC 地址两个字段组成，3 台交换机的网桥优先级值是默认值 32768。通过查看交换机 S1、S2 和 S3 的生成树配置可以发现，网桥优先级值是 32769。交换机默认启用的生成树协议是 PVST+，BID 由网桥优先级、扩展系统 ID 和 MAC 地址 3 部分组成。交换机使用的是默认 VLAN，只有 VLAN 1，扩展系统 ID 为 1，所以网桥优先级值是 32768+1=32769。交换机 S1、S2 和 S3 的网桥优先级值相同，所以重点比较 MAC 地址这个字段，因为交换机 S1 的 MAC 地址是 18ef.6394.3680，交换机 S2 的 MAC 地址是 18ef.638b.0500，交换机 S3 的 MAC 地址是 18ef.6366.7b80，3 台交换机中 MAC 地址数值最小的是交换机 S3，所以交换机 S3 当选为根桥。

（2）根端口的选举

在每个非根桥上都有唯一的一个根端口。交换机 S3 为根桥，交换机 S1 和交换机 S2 为非根桥，交换机 S1 和交换机 S2 上都有一个根端口，根端口的选举由非根桥到根桥之间的路径开销决定。因为 3 台交换机连接的都是 100 Mbit/s 的以太网口，这样交换机 S1 的 f0/1 端口到交换机 S3 的路径开销是 38，而交换机 S1 的 f0/2 端口到交换机 S3 的路径开销为 19，所以交换机 S1 的 f0/2 端口是根端口；交换机 S2 的 f0/1 端口到交换机 S3 的路径开销是 38，而交换机 S2 的 f0/2 端口到交换机 S3 的路径开销为 19，所以交换机 S2 的 f0/2 端口是根端口。

（3）指定端口和非指定端口的选举

每个网段都有唯一的一个指定端口，指定端口是到达根桥路径开销最小的端口。根桥上所有激活的端口都是指定端口，所以交换机 S3 的 f0/1 端口和 f0/2 端口都是指定端口。交换机 S1 和交换机 S2

之间的网段会有一个指定端口，可以发现，交换机 S1 的 f0/1 端口到根桥交换机 S3 的路径开销是 19，交换机 S2 的 f0/1 端口到根桥交换机 S3 的路径开销也是 19，两个端口到根桥的路径开销相同。这种情况下，比较 BID 值，若 BID 值中 3 台交换机的网桥优先级值相同，就重点比较 MAC 地址这个字段。交换机 S1 收到交换机 S2 发来的 BID 值，将其与自己的 BID 值进行比较，发现收到的交换机 S2 发来的 BID 值中的 MAC 地址是 18ef.638b.0500，其数值比自己的小，这样交换机 S2 的 f0/1 端口当选为指定端口，交换机 S1 的 f0/1 端口是非指定端口，该端口将处于阻塞状态。

16.4.2 STP 的配置

本小节网络拓扑如图 16-1 所示，要求完成 STP 的配置，使交换机 S1 成为根桥，使交换机 S2 的 f0/2 端口成为指定端口。

步骤 1：在交换机 S1 的全局配置模式下配置 S1 交换机为根桥，输入以下代码。

```
S1(config)#spanning-tree vlan 1 root primary
```

步骤 2：在交换机 S2 的全局配置模式下配置 S2 交换机为备用根桥，输入以下代码。

```
S2(config)#spanning-tree vlan 1 root secondary
```

步骤 3：在交换机 S1 的特权执行模式下输入 show spanning-tree，查看生成树的配置，如图 16-5 所示。

```
S1#show spanning-tree
VLAN0001
  Spanning tree enabled protocol ieee
  Root ID    Priority    24577
             Address     18ef.6394.3680
             This bridge is the root
             Hello Time   2 sec  Max Age 20 sec  Forward Delay 15 sec

  Bridge ID  Priority    24577   (priority 24576 sys-id-ext 1)
             Address     18ef.6394.3680
             Hello Time   2 sec  Max Age 20 sec  Forward Delay 15 sec
             Aging Time  300

Interface        Role Sts Cost      Prio.Nbr Type
---------------- ---- --- --------- -------- --------------------------------
Fa0/1            Desg FWD 19        128.1    P2p
Fa0/2            Desg FWD 19        128.2    P2p
Fa0/3            Desg FWD 19        128.3    P2p
```

图 16-5 交换机 S1 生成树的配置

步骤 4：在交换机 S2 的特权执行模式下输入 show spanning-tree，查看生成树的配置，如图 16-6 所示。

```
S2#show spanning-tree
VLAN0001
  Spanning tree enabled protocol ieee
  Root ID    Priority    24577
             Address     18ef.6394.3680
             Cost        19
             Port        1 (FastEthernet0/1)
             Hello Time   2 sec  Max Age 20 sec  Forward Delay 15 sec

  Bridge ID  Priority    28673   (priority 28672 sys-id-ext 1)
             Address     18ef.638b.0500
             Hello Time   2 sec  Max Age 20 sec  Forward Delay 15 sec
             Aging Time  300

Interface        Role Sts Cost      Prio.Nbr Type
---------------- ---- --- --------- -------- --------------------------------
Fa0/1            Root FWD 19        128.1    P2p
Fa0/2            Desg FWD 19        128.2    P2p
Fa0/3            Desg FWD 19        128.3    P2p
```

图 16-6 交换机 S2 生成树的配置

步骤 5：在交换机 S3 的特权执行模式下输入 show spanning-tree，查看生成树的配置，如图 16-7 所示。

```
S3#show spanning-tree
VLAN0001
  Spanning tree enabled protocol ieee
  Root ID    Priority    24577
             Address     18ef.6394.3680
             Cost        19
             Port        1 (FastEthernet0/1)
             Hello Time  2 sec  Max Age 20 sec  Forward Delay 15 sec

  Bridge ID  Priority    32769  (priority 32768 sys-id-ext 1)
             Address     18ef.6366.7b80
             Hello Time  2 sec  Max Age 20 sec  Forward Delay 15 sec
             Aging Time 300

Interface           Role Sts Cost      Prio.Nbr Type
------------------- ---- --- --------- -------- --------------------
Fa0/1               Root FWD 19        128.1    P2p
Fa0/2               Altn BLK 19        128.2    P2p
Fa0/3               Desg FWD 19        128.3    P2p
```

图 16-7　交换机 S3 生成树的配置

本小节通过 spanning-tree vlan 1 root primary 命令把交换机 S1 的网桥优先级值修改为 24577，使交换机 S1 成为根桥。通过分析可以发现，交换机 S2 的 f0/2 端口和交换机 S3 的 f0/2 端口到根桥的路径开销相同。这种情况下，比较 BID 值，若 BID 值中 3 台交换机的网桥优先级值相同，则重点比较 MAC 地址这个字段。交换机 S2 收到交换机 S3 发来的 BID 值，将其与自己的 BID 值进行比较，发现收到的交换机 S3 发来的 BID 值中的 MAC 地址是 18ef.6366.7b80，其数值比自己的小，这样交换机 S3 的 f0/2 端口当选为指定端口，交换机 S2 的 f0/2 端口为非指定端口。本小节要求交换机 S2 的 f0/2 端口为指定端口，通过 spanning-tree vlan 1 root secondary 命令把交换机 S2 的网桥优先级值修改为 28673，这样交换机 S2 的 BID 值小于交换机 S3 的 BID 值。通过查看生成树的配置，我们发现，交换机 S1 成为根桥，交换机 S2 的 f0/2 端口为指定端口。

16.4.3　边缘端口和 BPDU 防护的配置

本小节网络拓扑如图 16-1 所示，要求完成边缘端口和 BPDU 防护的配置，把交换机连接计算机的端口配置为边缘端口，并进行 BPDU 防护的配置。

步骤 1：在交换机 S1 的全局配置模式下配置 f0/3 端口为边缘端口，输入以下代码。

```
S1(config)#interface f0/3
S1(config-if)#switchport mode access
S1(config-if)#spanning-tree portfast
```

步骤 2：在交换机 S2 的全局配置模式下配置 f0/3 端口为边缘端口，输入以下代码。

```
S2(config)#interface f0/3
S2(config-if)#switchport mode access
S2(config-if)#spanning-tree portfast
```

步骤 3：在交换机 S3 的全局配置模式下配置 f0/3 端口为边缘端口，输入以下代码。

```
S3(config)#interface f0/3
S3(config-if)#switchport mode access
S3(config-if)#spanning-tree portfast
```

步骤 4：在交换机 S1 的全局配置模式下开启 BPDU 防护，输入以下代码。

```
S1(config)#spanning-tree portfast bpduguard default
```

步骤 5：在交换机 S2 的全局配置模式下开启 BPDU 防护，输入以下代码。

```
S2(config)#spanning-tree portfast bpduguard default
```

步骤 6：在交换机 S3 的全局配置模式下开启 BPDU 防护，输入以下代码。

```
S3(config)#spanning-tree portfast bpduguard default
```

步骤 7：把交换机 S1 的 f0/3 端口接入交换机 S2 的非边缘端口，会出现图 16-8 所示的提示信息。

```
*Mar  1 00:36:42.781: %SPANTREE-2-BLOCK_BPDUGUARD: Received BPDU on port Fa0/3 with BPDU Guard enabled. Disabling port.
*Mar  1 00:36:42.781: %PM-4-ERR_DISABLE: bpduguard error detected on Fa0/3, putting Fa0/3 in err-disable state
```

图 16-8 提示信息

步骤 8：在交换机 S1 的特权执行模式下输入 show int f0/3，查看交换机 S1 的 f0/3 端口的状态，如图 16-9 所示。

```
S1#show int f0/3
FastEthernet0/3 is down, line protocol is down (err-disabled)
  Hardware is Fast Ethernet, address is 18ef.6394.3683 (bia 18ef.6394.3683)
  MTU 1500 bytes, BW 10000 Kbit, DLY 1000 usec,
     reliability 255/255, txload 1/255, rxload 1/255
  Encapsulation ARPA, loopback not set
  Keepalive set (10 sec)
  Auto-duplex, Auto-speed, media type is 10/100BaseTX
  input flow-control is off, output flow-control is unsupported
  ARP type: ARPA, ARP Timeout 04:00:00
  Last input 00:00:27, output 00:00:27, output hang never
  Last clearing of "show interface" counters never
  Input queue: 0/75/0/0 (size/max/drops/flushes); Total output drops: 0
  Queueing strategy: fifo
  Output queue: 0/40 (size/max)
  5 minute input rate 0 bits/sec, 0 packets/sec
  5 minute output rate 0 bits/sec, 0 packets/sec
     1521 packets input, 131673 bytes, 0 no buffer
     Received 1521 broadcasts (657 multicasts)
     0 runts, 0 giants, 0 throttles
     0 input errors, 0 CRC, 0 frame, 0 overrun, 0 ignored
     0 watchdog, 657 multicast, 0 pause input
     0 input packets with dribble condition detected
     6416 packets output, 443333 bytes, 0 underruns
     0 output errors, 0 collisions, 1 interface resets
     0 babbles, 0 late collision, 0 deferred
     0 lost carrier, 0 no carrier, 0 PAUSE output
     0 output buffer failures, 0 output buffers swapped out
```

图 16-9 交换机 S1 的 f0/3 端口的状态

本小节把交换机 S1、S2 和 S3 连接终端的端口配置成了边缘端口，同时启用了 BPDU 防护，然后将交换机 S1 中启用了 BPDU 防护的 f0/3 端口连接交换机 S2 的非边缘端口，通过查看提示信息和查看端口状态可以发现，当交换机 S1 的 f0/3 边缘端口收到 BPDU 报文时，该端口会进入 error-disabled 状态。

16.5 项目小结

网络冗余提高了网络的可靠性，与此同时容易带来二层环路，从而在网络中产生大量无用的流量，引发广播风暴，造成网络崩溃。STP 可以通过计算阻塞某个端口以消除环路，将网络中的无用流量消除掉。当网络拓扑发生变化时，如某个转发接口连接的链路出现了故障，STP 就会通过计算使原来处于阻塞状态的端口进入转发状态，维持网络通信。为了使交换机连接终端的端口能够快速进入转发状态，我们可以把连接终端设备的交换端口配置成边缘端口，同时在边缘端口上开启 BPDU 防护，防止这个端口连接交换机。如果边缘端口收到了 BPDU 报文，则说明该端口连接了交换设备，有可能给网络带来环路，这时端口进入 error-disabled 状态，这将有效关闭端口。默认情况下，思科交换机的 STP 是开启的。

16.6 拓展训练

本项目的拓展训练网络拓扑如图 16-10 所示，要求完成如下配置：

（1）完成 3 台交换机的 STP 配置，使交换机 S1 当选为根桥，交换机 S2 当选为备份根桥，交换机 S3 的 f0/2 端口为指定端口；

（2）配置交换机 S3 连接终端设备的端口为边缘端口，并配置 BPDU 防护；

（3）查看交换机 S1、S2 和 S3 的 STP 配置，并查看根桥、根端口、指定端口和非指定端口。

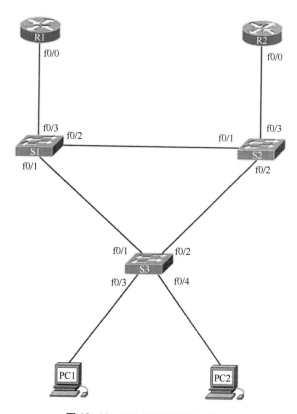

图 16-10 项目 16 拓展训练网络拓扑

项目 17

链路聚合的配置

17.1 用户需求

某学校网络拓扑如图 17-1 所示。交换机 S1 和 S2 互联的链路需要较大的带宽，那么怎样用比较经济的方式增大交换机 S1 和 S2 互联链路的带宽？

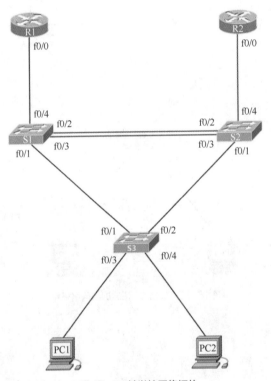

图 17-1 某学校网络拓扑

17.2 知识梳理

链路聚合

17.2.1 链路聚合

链路聚合能够使用两台设备之间的多个物理链路创建一条逻辑链路，这样，物理链路之间能够进行负载共享，而不是通过 STP 来阻塞一条或多条链路。以太网通道（EtherChannel）是交换网络中所使用的一种链路聚合形式。

17.2.2　EtherChannel 技术

EtherChannel 技术可以将多个快速以太网端口或吉比特以太网端口聚合到一个逻辑通道中，所产生的虚拟端口称为端口通道。物理端口捆绑在一起形成一个端口通道接口。

1. EtherChannel 技术的优势

（1）EtherChannel 技术依赖现有交换机端口，无须将链路升级为带宽更大，速度更快，价格更昂贵的链路。

（2）可以在端口通道接口下完成绝大多数配置任务，不需要对交换机的每个端口进行配置，可以确保通道中各条链路配置的一致性。

（3）同一个 EtherChannel 的不同链路之间可以进行负载均衡。可以根据硬件平台实际情况实施一个或多个负载均衡方法，包括物理链路上源 MAC 地址到目的 MAC 地址的负载均衡或源 IP 地址到目的 IP 地址的负载均衡。

（4）EtherChannel 可以实现冗余，创建的聚合被视为一条逻辑链路。因此，其中的一条物理链路断开并不会给拓扑带来变化，只要交换机之间有一条物理链路是正常工作的，EtherChannel 就会照常工作。

2. EtherChannel 技术的特点

（1）EtherChannel 能够将多个物理端口组合成一条或多条逻辑链路。

（2）EtherChannel 中端口的类型必须一致。

（3）EtherChannel 可提供 800 Mbit/s（快速以太网通道）或者 8 Gbit/s（吉比特以太网通道）的全双工带宽。

（4）EtherChannel 最多可由 8 个物理端口组合配置而成。

17.2.3　创建 EtherChannel 的两种协议

1. 端口聚合协议

端口聚合协议（Port Aggregation Protocol，PAgP）是思科的私有协议，可以用来自动创建 EtherChannel 链路。使用 PAgP 配置 EtherChannel 链路时，将在 EtherChannel 可用的端口之间发送 PAgP 数据包，以协商信道的形成。当 PAgP 识别到匹配的 EtherChannel 链路时，就将其分到同一 EtherChannel，然后作为单个端口通道接口被添加到生成树。

每 30s 发送一次 PAgP 数据包，PAgP 检查配置的一致性，并管理链路的添加，处理两台交换机之间的故障，确保创建 EtherChannel 时所有端口都具有相同类型的配置，即所有端口都必须具有相同的速度、双工设置和 VLAN 信息。创建通道后，修改任何端口都将改变其他所有端口。PAgP 通过检测两端的配置并确保链路的兼容性来协助创建 EtherChannel 链路，以便在需要时启用 EtherChannel 链路。PAgP 有以下 3 种模式。

（1）打开模式

打开（On）模式强制端口形成 EtherChannel，并且不使用 PAgP 进入通道。在打开模式下配置的端口不交换 PAgP 数据包。

（2）期望模式

期望（Desirable）模式将端口置于主动协商状态。在该状态下，端口通过发送 PAgP 数据包来发起与其他端口的协商。

（3）自动模式

自动（Auto）模式将端口置于被动协商状态。在该状态下，端口会响应它接收的 PAgP 数据包，但不会发起 PAgP 协商。

要形成 EtherChannel，两端的模式必须兼容。如果将一端配置为自动模式，那么只能等待另一端发起 EtherChannel 协商，此时将另一端配置为自动模式或者打开模式，将不能形成 EtherChannel。如果将一端配置为期望模式，就会主动发起 EtherChannel 协商，此时将另一端配置为期望模式或者自动模式都可以形成 EtherChannel。打开模式会将端口手动放置到 EtherChannel 中，不进行协商，只有将另一端也设置为打开模式才会形成 EtherChannel。

2. 链路聚合控制协议

链路聚合控制协议（Link Aggregation Control Protocol，LACP）允许将多个物理端口捆绑来形成单个逻辑通道，允许交换机通过向对等体发送 LACP 数据包协商自动捆绑。由于 LACP 是 IEEE 标准，所以可以在多供应商环境中使用，为 EtherChannel 提供便利。LACP 有以下 3 种模式。

（1）打开模式

打开（On）模式强制端口形成 EtherChannel，并且不使用 LACP 进入通道。打开模式下配置的端口不交换 LACP 数据包，不进行协商。

（2）主动模式

主动（Active）模式将端口置于主动协商状态。在该状态下，端口通过发送 LACP 数据包发起与其他端口的协商。

（3）被动模式

被动（Passive）模式将端口置于被动协商状态。在该状态下，端口会响应它接收的 LACP 数据包，但不会发起 LACP 协商。

像 PAgP 一样，两端的模式必须兼容才能形成 EtherChannel 链路。如果将一端配置为被动模式，那么只能等待另一端发起 EtherChannel 协商，此时将另一端设置为被动模式或者打开模式，将不能形成 EtherChannel。如果将一端配置为主动模式，就会主动发起 EtherChannel 协商，此时将另一端配置为主动模式或者被动模式都可以形成 EtherChannel。打开模式会无条件创建 EtherChannel 配置，无须使用 PAgP 或 LACP 动态协商。

17.2.4　配置原则

（1）所有模块上的所有以太网端口都必须支持 EtherChannel，不要求端口在物理上连续或位于同一模块。

（2）EtherChannel 中的所有端口都要以相同速度在相同双工模式下运行。

（3）为 EtherChannel 中的所有端口分配相同的 VLAN，或将其配置为 trunk 模式。

（4）在中继 EtherChannel 中的所有端口上，EtherChannel 都支持相同的 VLAN 允许范围。如果 VLAN 的允许范围不同，那么即使将其设置为自动或期望模式，端口也不会形成 EtherChannel。

17.2.5　配置命令

1. 指定构成 EtherChannel 组的端口

```
Switch(config)#interface range port-range
```

range：允许选择多个端口并将它们一起配置。

port-range：端口编号的范围。

2. 创建端口通道接口

```
Switch(config-if)#channel-group identifier mode active
```

identifier：标识符，指定 EtherChannel 组编号。

mode active：表示该接口会主动参与 LACP 的 EtherChannel 协商过程。

3. 进入端口通道接口配置模式

```
Switch(config-if)#interface port-channel identifier
```

identifier：标识符，指定通道组编号。

进入端口接口配置模式后，可以更改端口通道接口上的第 2 层设置。

4. 显示端口通道接口的总体状态

```
Switch#show interfaces port-channel
```

5. 用列表的每行显示一条端口通道接口信息

```
Switch#show etherchannel summary
```

6. 显示特定端口通道接口的信息

```
Switch#show etherchannel port-channel
```

7. 显示 EtherChannel 内每个端口的相关信息

```
Switch#show interfaces etherchannel
```

17.3 方案设计

在图 17-1 所示的网络拓扑中，要满足交换机 S1 和 S2 互联的链路具有较大带宽的需求，可以采用链路聚合的方式。这种方式依赖现有交换机端口，无须升级链路，所以采用这种方式既能增大带宽，也比较经济。

17.4 项目实施

本节网络拓扑如图 17-1 所示，交换机 S1 和 S2 通过两条线缆互联，要求配置 LACP，使交换机 S1 和 S2 互联的链路形成二层的 EtherChannel。

步骤 1：在交换机 S1 的全局配置模式下配置 EtherChannel，输入以下代码。

```
S1(config)#interface range f0/2-3
S1(config-if-range)#channel-group 1 mode active
```

步骤 2：在交换机 S2 的全局配置模式下配置 EtherChannel，输入以下代码。

```
S2(config)#interface range f0/2-3
S2(config-if-range)#channel-group 1 mode active
```

步骤 3：在交换机 S1 的特权执行模式下输入 show interfaces port-channel 1，查看端口通道接口的总体状态，如图 17-2 所示。

步骤 4：在交换机 S1 的特权执行模式下输入 show etherchannel summary，查看特定端口通道接口信息，如图 17-3 所示。

步骤 5：在交换机 S1 的特权执行模式下输入 show etherchannel port-channel，查看端口通道接口信息，如图 17-4 所示。

```
S1#show interfaces port-channel 1
Port-channel1 is up, line protocol is up (connected)
  Hardware is EtherChannel, address is 18ef.6366.7b82 (bia 18ef.6366.7b82)
  MTU 1500 bytes, BW 200000 Kbit, DLY 100 usec,
     reliability 255/255, txload 1/255, rxload 1/255
  Encapsulation ARPA, loopback not set
  Keepalive set (10 sec)
  Full-duplex, 100Mb/s, link type is auto, media type is unknown
  input flow-control is off, output flow-control is unsupported
  Members in this channel: Fa0/2 Fa0/3
  ARP type: ARPA, ARP Timeout 04:00:00
  Last input 00:00:25, output 00:00:01, output hang never
  Last clearing of "show interface" counters never
  Input queue: 0/75/0/0 (size/max/drops/flushes); Total output drops: 0
  Queueing strategy: fifo
  Output queue: 0/40 (size/max)
  5 minute input rate 0 bits/sec, 0 packets/sec
  5 minute output rate 2000 bits/sec, 2 packets/sec
     42 packets input, 5301 bytes, 0 no buffer
     Received 30 broadcasts (30 multicasts)
     0 runts, 0 giants, 0 throttles
     0 input errors, 0 CRC, 0 frame, 0 overrun, 0 ignored
     0 watchdog, 30 multicast, 0 pause input
     0 input packets with dribble condition detected
     130 packets output, 14902 bytes, 0 underruns
     0 output errors, 0 collisions, 1 interface resets
     0 babbles, 0 late collision, 0 deferred
     0 lost carrier, 0 no carrier, 0 PAUSE output
     0 output buffer failures, 0 output buffers swapped out
```

图 17-2　查看端口通道接口的总体状态

```
S1#show etherchannel summary
Flags:  D - down         P - bundled in port-channel
        I - stand-alone  s - suspended
        H - Hot-standby (LACP only)
        R - Layer3       S - Layer2
        U - in use       f - failed to allocate aggregator

        M - not in use, minimum links not met
        u - unsuitable for bundling
        w - waiting to be aggregated
        d - default port


Number of channel-groups in use: 1
Number of aggregators:           1

Group  Port-channel  Protocol    Ports
------+-------------+-----------+-----------------------------------------------
1      Po1(SU)         LACP      Fa0/2(P)   Fa0/3(P)
```

图 17-3　查看特定端口通道接口信息

```
S1#show etherchannel port-channel
                Channel-group listing:
                ----------------------

Group: 1
----------
                Port-channels in the group:
                ---------------------------

Port-channel: Po1    (Primary Aggregator)

------------

Age of the Port-channel   = 0d:00h:04m:08s
Logical slot/port   = 2/1       Number of ports = 2
HotStandBy port = null
Port state          = Port-channel Ag-Inuse
Protocol            =    LACP
Port security       = Disabled

Ports in the Port-channel:

Index   Load   Port     EC state              No of bits
------+------+--------+---------------------+----------
  0      00    Fa0/2    Active                  0
  0      00    Fa0/3    Active                  0

Time since last port bundled:    0d:00h:03m:21s    Fa0/3
```

图 17-4　查看端口通道接口信息

步骤 6：在交换机 S1 的特权执行模式下输入 show interfaces etherchannel，查看每个端口的相关信息，如图 17-5 所示。

```
S1#show interfaces etherchannel
----
FastEthernet0/2:
Port state    = Up Mstr Assoc In-Bndl
Channel group = 1            Mode = Active        Gcchange = -
Port-channel  = Po1           GC   = -            Pseudo port-channel = Po1
Port index    = 0            Load = 0x00          Protocol  =   LACP

Flags:  S - Device is sending Slow LACPDUs    F - Device is sending fast LACPDUs.
        A - Device is in active mode.         P - Device is in passive mode.

Local information:
                              LACP port   Admin   Oper   Port      Port
Port     Flags    State       Priority    Key     Key    Number    State
Fa0/2    SA       bndl        32768       0x1     0x1    0x2       0x3D

Partner's information:
                 LACP port                        Admin   Oper   Port      Port
Port     Flags   Priority    Dev ID       Age     key     Key    Number    State
Fa0/2    SA      32768       18ef.638b.0500  13s  0x0     0x1    0x2       0x3D

Age of the port in the current state: 0d:00h:05m:18s

----
FastEthernet0/3:
Port state    = Up Mstr Assoc In-Bndl
Channel group = 1            Mode = Active        Gcchange = -
Port-channel  = Po1           GC   = -            Pseudo port-channel = Po1
Port index    = 0            Load = 0x00          Protocol  =   LACP

Flags:  S - Device is sending Slow LACPDUs    F - Device is sending fast LACPDUs.
        A - Device is in active mode.         P - Device is in passive mode.

Local information:
                              LACP port   Admin   Oper   Port      Port
Port     Flags    State       Priority    Key     Key    Number    State
Fa0/3    SA       bndl        32768       0x1     0x1    0x3       0x3D

Partner's information:
                 LACP port                        Admin   Oper   Port      Port
Port     Flags   Priority    Dev ID       Age     key     Key    Number    State
Fa0/3    SA      32768       18ef.638b.0500  20s  0x0     0x1    0x3       0x3D

Age of the port in the current state: 0d:00h:05m:20s

----
Port-channel1:Port-channel1    (Primary aggregator)

Age of the Port-channel = 0d:00h:06m:10s
Logical slot/port   = 2/1           Number of ports = 2
HotStandBy port = null
Port state          = Port-channel Ag-Inuse
Protocol            =   LACP
Port security       = Disabled

Ports in the Port-channel:

Index  Load   Port     EC state          No of bits
------+------+------+------------------+-----------
0      00     Fa0/2    Active            0
0      00     Fa0/3    Active            0

Time since last port bundled:    0d:00h:05m:23s    Fa0/3
```

图 17-5　查看每个端口的相关信息

步骤 7：在交换机 S2 的特权执行模式下输入 show interfaces port-channel 1，查看端口通道接口的总体状态，如图 17-6 所示。

```
S2#show interfaces port-channel 1
Port-channel1 is up, line protocol is up (connected)
  Hardware is EtherChannel, address is 18ef.638b.0502 (bia 18ef.638b.0502)
  MTU 1500 bytes, BW 200000 kbit, DLY 100 usec,
     reliability 255/255, txload 1/255, rxload 1/255
  Encapsulation ARPA, loopback not set
  Keepalive set (10 sec)
  Full-duplex, 100Mb/s, link type is auto, media type is unknown
  input flow-control is off, output flow-control is unsupported
  Members in this channel: Fa0/2 Fa0/3
  ARP type: ARPA, ARP Timeout 04:00:00
  Last input 00:00:01, output 00:08:12, output hang never
  Last clearing of "show interface" counters never
  Input queue: 0/75/0/0 (size/max/drops/flushes); Total output drops: 0
  Queueing strategy: fifo
  Output queue: 0/40 (size/max)
  5 minute input rate 1000 bits/sec, 1 packets/sec
  5 minute output rate 0 bits/sec, 0 packets/sec
     785 packets input, 111594 bytes, 0 no buffer
     Received 681 broadcasts (452 multicasts)
     0 runts, 0 giants, 0 throttles
     0 input errors, 0 CRC, 0 frame, 0 overrun, 0 ignored
     0 watchdog, 453 multicast, 0 pause input
     0 input packets with dribble condition detected
     243 packets output, 25975 bytes, 0 underruns
     0 output errors, 0 collisions, 1 interface resets
     0 babbles, 0 late collision, 0 deferred
     0 lost carrier, 0 no carrier, 0 PAUSE output
     0 output buffer failures, 0 output buffers swapped out
```

图 17-6　查看端口通道接口的总体状态

步骤 8：在交换机 S2 的特权执行模式下输入 show etherchannel summary，查看特定端口通道接口信息，如图 17-7 所示。

步骤 9：在交换机 S2 的特权执行模式下输入 show etherchannel port-channel，查看端口通道接口信息，如图 17-8 所示。

图 17-7　查看特定端口通道接口信息

图 17-8　查看端口通道接口信息

步骤 10：在交换机 S2 的特权执行模式下输入 show interfaces etherchannel，查看每个端口的相关信息，如图 17-9 所示。

图 17-9　查看每个端口的相关信息

通过查看端口通道接口的状态可以发现，端口通道 1 接口已经启用。通过查看端口通道接口信息和组成端口通道的物理端口的信息可以发现，交换机配置了一个 EtherChannel，EtherChannel 组 1 使用 LACP，端口通道 1 接口由两个物理端口（f0/2 和 f0/3）组成，交换机 S1 和 S2 都在主动模式下使用 LACP 成功建立 EtherChannel。在命令 show etherchannel summary 的执行结果中，端口通道编号旁边的字母"SU"表示第 2 层 EtherChannel 组正在使用。

17.5 项目小结

本项目完成了链路聚合的配置，EtherChannel 是一种链路聚合形式。创建 EtherChannel 的协议有 PAgP 和 LACP 两种，其中，PAgP 是思科的私有协议，LACP 是 IEEE 标准。使用这两种协议都可以形成 EtherChannel 链路，把交换机的多个端口聚合起来，可以增大链路带宽，提高链路的可靠性。

17.6 拓展训练

本项目的拓展训练网络拓扑如图 17-10 所示，要求完成链路聚合的配置：
（1）用 PAgP 使交换机 S1 和 S2 互联的链路形成 EtherChannel 链路；
（2）用 LACP 使交换机 S1 和 S3 互联的链路形成 EtherChannel 链路；
（3）强制交换机 S2 和 S3 互联的链路形成 EtherChannel 链路。

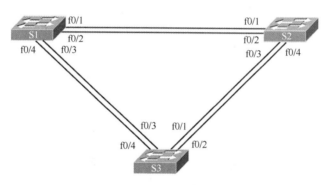

图 17-10　项目 17 拓展训练网络拓扑

项目 18

交换机端口安全的配置

18.1 用户需求

某学校办公网络部分的网络拓扑如图 18-1 所示。为了提高网络的安全性，让接入交换机的不合法用户不能访问网络，怎样实现这样的功能？

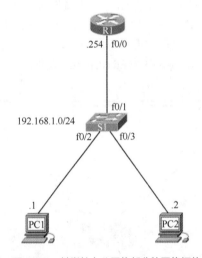

图 18-1 某学校办公网络部分的网络拓扑

18.2 知识梳理

18.2.1 端口安全

端口安全分为静态端口安全、动态端口安全和粘滞端口安全 3 种。端口安全限制端口上所允许的有效 MAC 地址的数量，可以为安全端口分配安全 MAC 地址，当数据包的源地址不是已定义的安全 MAC 地址时，端口不会转发该数据包。如果将安全 MAC 地址的数量限制为一个，并只为该端口分配一个安全 MAC 地址，那么只有地址为该特定安全 MAC 地址的工作站才能成功连接到该交换机端口，而且连接该端口的工作站将确保获得端口的全部带宽。如果交换机端口的安全 MAC 地址的数量已达到最大值，那么当尝试访问该端口的工作站的 MAC 地址不同于任何已确定的安全 MAC

端口安全

地址时，会发生安全违规。

> 🔍 **边学边思**
>
> 网络安全违规也是违法行为。我们要树立遵纪守法的意识，认真遵守各项纪律、规章和制度，做一个合法用户，努力为法治中国建设贡献力量。
>
> ① 学习法律知识：了解并学习相关的法律法规，包括《中华人民共和国宪法》《中华人民共和国刑法》《中华人民共和国行政处罚法》等。这有助于明确自己的权利和义务，并为合法行为提供指引。
>
> ② 遵守法律法规：遵循国家的法律法规，不进行任何违法犯罪行为。尊重他人的合法权益，不侵犯他人的人身、财产等权益。
>
> ③ 遵守社会规范：积极遵守社会公共秩序和道德规范，遵守社会公德、职业道德和行业规范。不参与传播虚假信息、恶意攻击他人等违背道德原则的活动。
>
> ④ 加强法律意识：提高对法律风险的警觉性，避免进行涉及法律风险的行为。如果遇到法律问题，要及时咨询专业律师或相关机构，以获得合法的解决方案。
>
> ⑤ 守信用、诚实守法：在经济、商业和社交活动中，要讲究诚信，履行承诺。遵守市场规则，不进行欺诈、虚假宣传等违法行为。
>
> ⑥ 积极参与公共事务：关注社会公共事务，积极参与社区、城市和国家的建设。通过参与公共讨论、履行公民义务、提供建设性的意见和建议，为法治中国的建设贡献力量。

18.2.2 安全 MAC 地址类型

安全 MAC 地址类型主要有静态安全 MAC 地址、动态安全 MAC 地址和粘滞安全 MAC 地址这 3 类。

1. 静态安全 MAC 地址

静态安全 MAC 地址是使用命令手动配置的交换机端口的安全 MAC 地址，存储在 MAC 地址表中，并被添加到交换机的运行配置中。

2. 动态安全 MAC 地址

动态安全 MAC 地址是动态学习的，并且仅存储在 MAC 地址表中。动态安全 MAC 地址在重新启动交换机时将被移除。

3. 粘滞安全 MAC 地址

当在端口上启用安全 MAC 地址粘滞学习时，端口将所有动态安全 MAC 地址（包括那些在启用安全 MAC 地址粘滞获取之前动态获得的 MAC 地址）转换为粘滞安全 MAC 地址，并将所有粘滞安全 MAC 地址添加到运行配置中。如果禁用端口安全 MAC 地址粘滞学习，则粘滞安全 MAC 地址仍作为 MAC 地址表的一部分，但会从运行配置中被移除。如果启用端口安全 MAC 地址粘滞学习，那么端口学习粘滞安全 MAC 地址后禁用端口安全，粘滞安全 MAC 地址仍保留在运行配置中。如果将粘滞安全 MAC 地址保存在配置文件中，则当交换机重新启动或者接口关闭时，接口不需要重新学习这些地址。

18.2.3 安全违规模式

安全违规模式主要有保护、限制和关闭 3 种。

1. 保护安全违规模式

保护安全违规模式下，当安全 MAC 地址的数量达到端口允许的数量时，带有未知源地址的数据包将被丢弃，直至移除地址使安全 MAC 地址的数量在允许的范围内，或增加允许的地址数量。保护安全违规模式不会发送安全违规的通知。

2. 限制安全违规模式

限制安全违规模式下，当安全 MAC 地址的数量达到端口允许的数量时，带有未知源地址的数据包将被丢弃，直至移除地址使安全 MAC 地址的数量在允许的范围内，或者增加允许的地址数量。该模式会发送 SNMP Trap 消息，记录系统日志，并增加违规计数器的计数。

3. 关闭安全违规模式

关闭安全违规模式下，端口安全违规将导致端口立即进入 error-disabled 状态，并关闭端口 LED。该模式还会发送 SNMP Trap 消息，记录系统日志，并增加违规计数器的计数。当端口处于 error-disabled 状态时，要恢复该端口，需要移除地址，使安全 MAC 地址的数量在允许的范围内，或增加允许的地址数量，然后用 shutdown 命令关闭端口，再用 no shutdown 命令开启端口。默认情况下，交换机端口的安全违规模式是关闭安全违规模式。

18.2.4 发生安全违规的条件

（1）在 MAC 地址表中添加了最大数量的安全 MAC 地址，还有工作站试图访问端口，且该工作站的 MAC 地址未出现在该 MAC 地址表中。

（2）在一个端口上学习或配置的地址出现在同一个 VLAN 中的另一个端口上。

18.2.5 配置命令

1. 启用端口安全和设置端口的安全违规模式

`Switch(config-if)#switchport port-security [violation {protect | restrict | shutdown}]`

switchport port-security：启用端口安全，只有交换机的静态接入端口或中继端口才能开启端口安全。

protect：将违规的 MAC 地址的流量丢弃，不记录。

restrict：将违规的 MAC 地址的流量丢弃，并发送 SNMP Trap 消息，记录系统日志。

shutdown：将安全违规模式设置为关闭安全违规模式，违规后执行动作 shutdown（error-disabled），并发送 SNMP Trap 消息，记录系统日志，是端口的默认安全违规模式。

2. 设置端口允许通过的最大 MAC 地址数量

`Switch(config-if)#switchport port-security maximum value`

value：端口允许通过的最大 MAC 地址数量。

3. 设置静态安全 MAC 地址

`Switch(config-if)#switchport port-security mac-address mac-address`

mac-address：接入端口设备的静态 MAC 地址。

4. 设置粘滞安全 MAC 地址

`Switch(config-if)#switchport port-security mac-address sticky { mac-address}`

5. 显示交换机或指定端口的安全设置

`Switch# show port-security [interface interface-id]`

6. 显示所有交换机端口或某个指定端口上配置的所有安全 MAC 地址

```
Switch# show port-security [interface interface-id] address
```

18.3　方案设计

在图 18-1 所示的网络拓扑中，要实现不允许接入交换机的不合法用户访问网络，可以启用端口安全，启用端口安全后，不合法的设备无法访问网络，可以为安全端口分配静态端口安全地址、动态端口安全地址或粘滞端口安全地址，可将端口的安全违规模式配置为保护安全违规模式、限制安全违规模式或关闭安全违规模式。

18.4　项目实施

18.4.1　静态端口安全的配置

本小节网络拓扑如图 18-1 所示，计算机 PC1 的 MAC 地址为 54be.f74e.2456，计算机 PC2 的 MAC 地址为 54be.f74e.2502。为了防止不合法用户访问网络，要求配置静态端口安全，使交换机的 f0/2 端口只允许计算机 PC1 接入，交换机的 f0/3 端口只允许计算机 PC2 接入，一旦两个端口上有非法计算机接入，就关闭端口。

步骤 1：在交换机 S1 的全局配置模式下启用端口安全，输入以下代码。

```
S1(config)#interface range f0/2-3
S1(config-if)#switchport mode access
S1(config-if)#switchport port-security
```

步骤 2：在交换机 S1 的全局配置模式下配置 f0/2 端口的静态安全地址，输入以下代码。

```
S1(config)#interface f0/2
S1(config-if)#switchport port-security mac-address 54be.f74e.2456
```

步骤 3：在交换机 S1 的全局配置模式下配置 f0/3 端口的静态安全地址，输入以下代码。

```
S1(config)#interface f0/3
S1(config-if)#switchport port-security mac-address 54be.f74e.2502
```

步骤 4：在交换机 S1 的特权执行模式下输入 show port-security address，查看安全 MAC 地址，如图 18-2 所示。

```
S1#show port-security address
        Secure Mac Address Table
-------------------------------------------------------------------
Vlan    Mac Address       Type              Ports   Remaining Age
                                                      (mins)
----    -----------       ----              -----   -------------
  1     54be.f74e.2456    SecureConfigured  Fa0/2       -
  1     54be.f74e.2502    SecureConfigured  Fa0/3       -
-------------------------------------------------------------------
Total Addresses in System (excluding one mac per port)     : 0
Max Addresses limit in System (excluding one mac per port) : 8192
```

图 18-2　交换机 S1 的安全 MAC 地址

步骤 5：在交换机 S1 的特权执行模式下输入 show port-security interface f0/2，查看 f0/2 端口的安全设置，如图 18-3 所示。

```
S1#show port-security interface f0/2
Port Security              : Enabled
Port Status                : Secure-up
Violation Mode             : Shutdown
Aging Time                 : 0 mins
Aging Type                 : Absolute
SecureStatic Address Aging : Disabled
Maximum MAC Addresses      : 1
Total MAC Addresses        : 1
Configured MAC Addresses   : 1
Sticky MAC Addresses       : 0
Last Source Address:Vlan   : 54be.f74e.2456:1
Security Violation Count   : 0
```

图 18-3 查看交换机 S1 的 f0/2 端口的安全设置

步骤 6：在交换机 S1 的特权执行模式下输入 show port-security interface f0/3，查看 f0/3 端口的安全设置，如图 18-4 所示。

```
S1#show port-security interface f0/3
Port Security              : Enabled
Port Status                : Secure-up
Violation Mode             : Shutdown
Aging Time                 : 0 mins
Aging Type                 : Absolute
SecureStatic Address Aging : Disabled
Maximum MAC Addresses      : 1
Total MAC Addresses        : 1
Configured MAC Addresses   : 1
Sticky MAC Addresses       : 0
Last Source Address:Vlan   : 54be.f74e.2502:1
Security Violation Count   : 0
```

图 18-4 查看交换机 S1 的 f0/3 端口的安全设置

步骤 7：在交换机 S1 的特权执行模式下输入 show run interface f0/2，查看交换机 S1 运行配置文件中 f0/2 端口的配置，如图 18-5 所示。

```
S1#show run interface f0/2
Building configuration...

Current configuration : 136 bytes
!
interface FastEthernet0/2
 switchport mode access
 switchport port-security
 switchport port-security mac-address 54be.f74e.2456
end
```

图 18-5 交换机 S1 运行配置文件中 f0/2 端口的配置

步骤 8：在交换机 S1 的特权执行模式下输入 show run interface f0/3，查看交换机 S1 运行配置文件中 f0/3 端口的配置，如图 18-6 所示。

```
S1#show run interface f0/3
Building configuration...

Current configuration : 136 bytes
!
interface FastEthernet0/3
 switchport mode access
 switchport port-security
 switchport port-security mac-address 54be.f74e.2502
end
```

图 18-6 交换机 S1 运行配置文件中 f0/3 端口的配置

步骤 9：在计算机 PC1 的命令行界面输入 ping 192.168.1.2，检验联通性，如图 18-7 所示。

项目 18
交换机端口安全的配置

```
C:\>ping 192.168.1.2

正在 Ping 192.168.1.2 具有 32 字节的数据:
来自 192.168.1.2 的回复: 字节=32 时间<1ms TTL=128
来自 192.168.1.2 的回复: 字节=32 时间<1ms TTL=128
来自 192.168.1.2 的回复: 字节=32 时间<1ms TTL=128
来自 192.168.1.2 的回复: 字节=32 时间<1ms TTL=128

192.168.1.2 的 Ping 统计信息:
    数据包: 已发送 = 4, 已接收 = 4, 丢失 = 0 (0% 丢失),
往返行程的估计时间(以毫秒为单位):
    最短 = 0ms, 最长 = 0ms, 平均 = 0ms
```

图 18-7 从计算机 PC1 ping 计算机 PC2

步骤 10:把连接计算机 PC1 和 PC2 的两个交换机端口互换,会显示图 18-8 所示的提示消息。

```
*Mar  1 00:41:51.876: %PORT_SECURITY-2-PSECURE_VIOLATION: Security violation occurred, caused by MAC address 54be.f74e.24
56 on port FastEthernet0/3.
*Mar  1 00:41:52.882: %LINEPROTO-5-UPDOWN: Line protocol on Interface FastEthernet0/3, changed state to down
*Mar  1 00:41:53.881: %LINK-3-UPDOWN: Interface FastEthernet0/3, changed state to down
*Mar  1 00:41:53.956: %PM-4-ERR_DISABLE: psecure-violation error detected on Fa0/2, putting Fa0/2 in err-disable state
*Mar  1 00:41:54.963: %LINEPROTO-5-UPDOWN: Line protocol on Interface FastEthernet0/2, changed state to down
*Mar  1 00:41:55.961: %LINK-3-UPDOWN: Interface FastEthernet0/2, changed state to down
```

图 18-8 提示消息

步骤 11:在交换机 S1 的特权执行模式下查看交换机 S1 的 f0/2 端口的状态,如图 18-9 所示。

```
S1#sh int f0/2 status

Port      Name              Status       Vlan    Duplex  Speed Type
Fa0/2                       err-disabled 1         auto   auto 10/100BaseTX
```

图 18-9 交换机 S1 的 f0/2 端口的状态

步骤 12:在交换机 S1 的特权执行模式下查看交换机 S1 的 f0/2 端口的安全设置,如图 18-10 所示。

```
S1#show port-security interface f0/2
Port Security                : Enabled
Port Status                  : Secure-shutdown
Violation Mode               : Shutdown
Aging Time                   : 0 mins
Aging Type                   : Absolute
SecureStatic Address Aging   : Disabled
Maximum MAC Addresses        : 1
Total MAC Addresses          : 1
Configured MAC Addresses     : 1
Sticky MAC Addresses         : 0
Last Source Address:Vlan     : 54be.f74e.2502:1
Security Violation Count     : 1
```

图 18-10 交换机 S1 的 f0/2 端口的安全设置

步骤 13:在交换机 S1 的特权执行模式下查看交换机 S1 的 f0/3 端口的状态,如图 18-11 所示。

```
S1#show interface f0/3 status

Port      Name              Status       Vlan    Duplex  Speed Type
Fa0/3                       err-disabled 1         auto   auto 10/100BaseTX
```

图 18-11 交换机 S1 的 f0/3 端口的状态

步骤 14:在交换机 S1 的特权执行模式下查看交换机 S1 的 f0/3 端口的安全设置,如图 18-12 所示。

```
S1#show port-security interface f0/3
Port Security                : Enabled
Port Status                  : Secure-shutdown
Violation Mode               : Shutdown
Aging Time                   : 0 mins
Aging Type                   : Absolute
SecureStatic Address Aging   : Disabled
Maximum MAC Addresses        : 1
Total MAC Addresses          : 1
Configured MAC Addresses     : 1
Sticky MAC Addresses         : 0
Last Source Address:Vlan     : 54be.f74e.2456:1
Security Violation Count     : 1
```

图 18-12 交换机 S1 的 f0/3 端口的安全设置

步骤 15：在计算机 PC1 的命令行界面输入 ping 192.168.1.2，检验联通性，如图 18-13 所示。

```
C:\>ping 192.168.1.2

正在 Ping 192.168.1.2 具有 32 字节的数据:
来自 192.168.1.1 的回复: 无法访问目标主机。
来自 192.168.1.1 的回复: 无法访问目标主机。
来自 192.168.1.1 的回复: 无法访问目标主机。
来自 192.168.1.1 的回复: 无法访问目标主机。

192.168.1.2 的 Ping 统计信息:
    数据包: 已发送 = 4，已接收 = 4，丢失 = 0 (0% 丢失)，
```

图 18-13　从计算机 PC1 ping 计算机 PC2

步骤 16：要重启端口，必须把违规的计算机从交换机端口移除，然后执行 shutdown 和 no shutdown 命令重启端口。

```
S1(config)#interface range f0/2-3
S1(config-if-range)#shutdown
S1(config-if-range)#no shutdown
```

重启端口后，发现显示图 18-14 所示的提示信息，说明交换机的端口正常启动了。

```
*Mar  1 00:27:20.803: %LINK-5-CHANGED: Interface FastEthernet0/2, changed state to administrativel
y down
*Mar  1 00:27:20.803: %LINK-5-CHANGED: Interface FastEthernet0/3, changed state to administrativel
y down
*Mar  1 00:27:23.621: %LINK-3-UPDOWN: Interface FastEthernet0/2, changed state to up
*Mar  1 00:27:23.630: %LINK-3-UPDOWN: Interface FastEthernet0/3, changed state to up
*Mar  1 00:27:24.628: %LINEPROTO-5-UPDOWN: Line protocol on Interface FastEthernet0/2, changed sta
te to up
*Mar  1 00:27:24.636: %LINEPROTO-5-UPDOWN: Line protocol on Interface FastEthernet0/3, changed sta
te to up
```

图 18-14　提示信息

通过查看交换机 S1 的 f0/2 端口和 f0/3 端口的状态与安全设置可以发现，当完成静态端口安全的配置后，交换机的静态安全 MAC 地址会被写入运行配置文件中；当把计算机 PC1 连接到交换机 S1 的 f0/3 端口，把计算机 PC2 连接到交换机 S1 的 f0/2 端口时，交换机 S1 的 f0/2 端口和 f0/3 端口处于 error-disabled 状态，会显示提示信息，违规计数器的数值会增加。本小节中，开启了使用命令配置端口的安全违规模式，默认情况下，端口安全违规模式是关闭的，若想使用需要手动开启。

18.4.2　动态端口安全的配置

本小节网络拓扑如图 18-1 所示。为了防止不合法的用户访问网络，要求配置动态端口安全，使交换机 S1 的 f0/3 端口只允许两台计算机接入，一旦交换机端口上有不合法的计算机接入，端口就进入保护安全违规模式。

步骤 1：在交换机 S1 的全局配置模式下启用端口安全，输入以下代码。

```
S1(config)#interface f0/3
S1(config-if)#switchport mode access
S1(config-if)#switchport port-security
```

步骤 2：配置端口合法的地址数量，输入以下代码。

```
S1(config-if)#switchport port-security maximum 2
```

步骤 3：配置端口的安全违规模式，输入以下代码。

```
S1(config-if)#switchport port-security violation protect
```

步骤 4：把一台集线器 HUB1 接入交换机 S1 的 f0/3 端口，并在集线器 HUB1 上接入两台计算机，网络拓扑如图 18-15 所示。

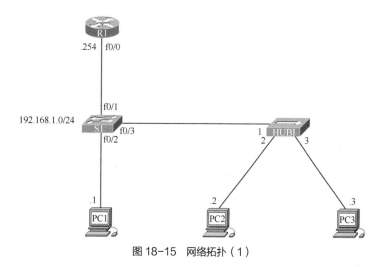

图 18-15　网络拓扑（1）

步骤 5：在交换机 S1 的特权执行模式下输入 show port-security address，查看安全 MAC 地址，如图 18-16 所示。

```
S1#show port-security address
        Secure Mac Address Table
-------------------------------------------------------------
Vlan    Mac Address       Type          Ports   Remaining Age
                                                   (mins)
----    -----------       ----          -----   -------------
  1     54be.f74e.2456    SecureDynamic  Fa0/3        -
  1     54be.f74e.2502    SecureDynamic  Fa0/3        -
-------------------------------------------------------------
Total Addresses in System (excluding one mac per port)     : 1
Max Addresses limit in System (excluding one mac per port) : 8192
```

图 18-16　交换机 S1 的安全 MAC 地址

步骤 6：在交换机 S1 的特权执行模式下输入 show port-security interface f0/3，查看 f0/3 端口的安全设置，如图 18-17 所示。

```
S1#show port-security interface f0/3
Port Security              : Enabled
Port Status                : Secure-up
Violation Mode             : Protect
Aging Time                 : 0 mins
Aging Type                 : Absolute
SecureStatic Address Aging : Disabled
Maximum MAC Addresses      : 2
Total MAC Addresses        : 2
Configured MAC Addresses   : 0
Sticky MAC Addresses       : 0
Last Source Address:Vlan   : 54be.f74e.2456:1
Security Violation Count   :
```

图 18-17　交换机 S1 的 f0/3 端口的安全设置

步骤 7：在计算机 PC2 的命令行界面输入 ping 192.168.1.254，检验联通性，如图 18-18 所示。

```
C:\>ping 192.168.1.254

正在 Ping 192.168.1.254 具有 32 字节的数据:
来自 192.168.1.254 的回复: 字节=32 时间=1ms TTL=255
来自 192.168.1.254 的回复: 字节=32 时间<1ms TTL=255
来自 192.168.1.254 的回复: 字节=32 时间<1ms TTL=255
来自 192.168.1.254 的回复: 字节=32 时间<1ms TTL=255

192.168.1.254 的 Ping 统计信息:
    数据包: 已发送 = 4，已接收 = 4，丢失 = 0 (0% 丢失)，
往返行程的估计时间(以毫秒为单位):
    最短 = 0ms，最长 = 1ms，平均 = 0ms
```

图 18-18　从计算机 PC2 ping 路由器 R1 的 f0/0 接口

步骤 8：在计算机 PC3 的命令行界面输入 ping 192.168.1.254，检验联通性，如图 18-19 所示。

```
C:\>ping 192.168.1.254

正在 Ping 192.168.1.254 具有 32 字节的数据：
来自 192.168.1.254 的回复: 字节=32 时间=2ms TTL=255
来自 192.168.1.254 的回复: 字节=32 时间<1ms TTL=255
来自 192.168.1.254 的回复: 字节=32 时间<1ms TTL=255
来自 192.168.1.254 的回复: 字节=32 时间<1ms TTL=255

192.168.1.254 的 Ping 统计信息：
    数据包: 已发送 = 4，已接收 = 4，丢失 = 0 (0% 丢失)，
往返行程的估计时间(以毫秒为单位)：
    最短 = 0ms，最长 = 2ms，平均 = 0ms
```

图 18-19　从计算机 PC3 ping 路由器 R1 的 f0/0 接口

步骤 9：在集线器 HUB1 上接入第三台计算机 PC4，网络拓扑如图 18-20 所示。

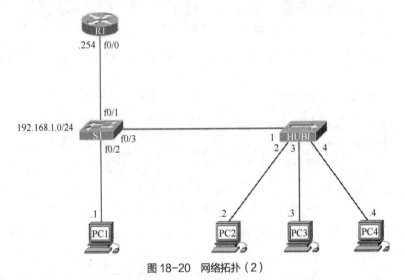

图 18-20　网络拓扑（2）

步骤 10：在交换机 S1 的特权执行模式下输入 show interface f0/3 status，查看交换机 S1 的 f0/3 端口的状态，如图 18-21 所示。

```
S1#show interface f0/3 status

Port      Name        Status        Vlan    Duplex  Speed Type
Fa0/3                 connected     1       a-full  a-100 10/100BaseTX
```

图 18-21　交换机 S1 的 f0/3 端口的状态

步骤 11：在交换机 S1 的特权执行模式下输入 show port-security interface f0/3，查看 f0/3 端口的安全设置，如图 18-22 所示。

```
S1#show port-security interface f0/3
Port Security                 : Enabled
Port Status                   : Secure-up
Violation Mode                : Protect
Aging Time                    : 0 mins
Aging Type                    : Absolute
SecureStatic Address Aging    : Disabled
Maximum MAC Addresses         : 2
Total MAC Addresses           : 2
Configured MAC Addresses      : 0
Sticky MAC Addresses          : 0
Last Source Address:Vlan      : 54be.f74e.2456:1
Security Violation Count      : 0
```

图 18-22　交换机 S1 的 f0/3 端口的安全设置

步骤 12：在交换机 S1 的特权执行模式下输入 show port-security address，查看安全 MAC 地址，如图 18-23 所示。

```
S1#show port-security address
          Secure Mac Address Table
-------------------------------------------------------------------
Vlan    Mac Address       Type              Ports   Remaining Age
                                                        (mins)
----    -----------       ----              -----   -------------
  1     54be.f74e.2456    SecureDynamic     Fa0/3        -
  1     54be.f74e.2502    SecureDynamic     Fa0/3        -
-------------------------------------------------------------------
Total Addresses in System (excluding one mac per port)       : 1
Max Addresses limit in System (excluding one mac per port)   : 8192
```

图 18-23　交换机 S1 的安全 MAC 地址

步骤 13：在计算机 PC4 的命令行界面输入 ping 192.168.1.2，检验联通性，如图 18-24 所示。

```
C:\>ping 192.168.1.2

正在 Ping 192.168.1.2 具有 32 字节的数据：
来自 192.168.1.2 的回复：字节=32 时间=1ms TTL=128
来自 192.168.1.2 的回复：字节=32 时间<1ms TTL=128
来自 192.168.1.2 的回复：字节=32 时间<1ms TTL=128
来自 192.168.1.2 的回复：字节=32 时间<1ms TTL=128

192.168.1.2 的 Ping 统计信息：
    数据包：已发送 = 4，已接收 = 4，丢失 = 0 (0% 丢失)，
往返行程的估计时间(以毫秒为单位)：
    最短 = 0ms，最长 = 1ms，平均 = 0ms
```

图 18-24　从计算机 PC4 ping 计算机 PC2

步骤 14：在计算机 PC4 的命令行界面输入 ping 192.168.1.3，检验联通性，如图 18-25 所示。

```
C:\>ping 192.168.1.3

正在 Ping 192.168.1.3 具有 32 字节的数据：
来自 192.168.1.3 的回复：字节=32 时间<1ms TTL=128
来自 192.168.1.3 的回复：字节=32 时间<1ms TTL=128
来自 192.168.1.3 的回复：字节=32 时间<1ms TTL=128
来自 192.168.1.3 的回复：字节=32 时间<1ms TTL=128

192.168.1.3 的 Ping 统计信息：
    已发送 = 4，已接收 = 4，丢失 = 0 (0% 丢失)，
往返行程的估计时间(以毫秒为单位)：
    最短 = 0ms，最长 = 0ms，平均 = 0ms
```

图 18-25　从计算机 PC4 ping 计算机 PC3

步骤 15：在计算机 PC4 的命令行界面输入 ping 192.168.1.254，检验联通性，如图 18-26 所示。

```
C:\>ping 192.168.1.254

正在 Ping 192.168.1.254 具有 32 字节的数据：
来自 192.168.1.4 的回复：无法访问目标主机。
来自 192.168.1.4 的回复：无法访问目标主机。
来自 192.168.1.4 的回复：无法访问目标主机。
来自 192.168.1.4 的回复：无法访问目标主机。

192.168.1.254 的 Ping 统计信息：
    数据包：已发送 = 4，已接收 = 4，丢失 = 0 (0% 丢失)，
```

图 18-26　从计算机 PC4 ping 路由器 R1 的 f0/0 接口

步骤 16：在计算机 PC3 的命令行界面输入 ping 192.168.1.254，检验联通性，如图 18-27 所示。

```
C:\>ping 192.168.1.254

正在 Ping 192.168.1.254 具有 32 字节的数据:
来自 192.168.1.254 的回复: 字节=32 时间=2ms TTL=255
来自 192.168.1.254 的回复: 字节=32 时间<1ms TTL=255
来自 192.168.1.254 的回复: 字节=32 时间<1ms TTL=255
来自 192.168.1.254 的回复: 字节=32 时间<1ms TTL=255

192.168.1.254 的 Ping 统计信息:
    数据包: 已发送 = 4, 已接收 = 4, 丢失 = 0 (0% 丢失),
往返行程的估计时间(以毫秒为单位):
    最短 = 0ms, 最长 = 2ms, 平均 = 0ms
```

图 18-27 从计算机 PC3 ping 路由器 R1 的 f0/0 接口

通过查看交换机 S1 的 f0/3 端口的状态和安全设置可以发现,当交换机 S1 的 f0/3 端口接入的计算机数量超过两台时,就出现了端口的安全违规,端口进入保护安全违规模式,不显示提示信息,违规计数器不计数。通过查看从计算机 PC4 ping 计算机 PC2 和 PC3 的结果可以发现,计算机 PC2、PC3 和 PC4 已经实现了互通,说明计算机 PC4 已经正确接入网络。通过查看从计算机 PC3 和计算机 PC4 ping 路由器 R1 的 f0/0 接口的结果可以发现,计算机 PC3 能够通过交换机 S1 的 f0/3 端口访问路由器 R1 的 f0/0 接口,而计算机 PC4 无法通过交换机 S1 的 f0/3 端口访问路由器 R1 的 f0/0 接口。保护安全违规模式下不会发送安全违规的通知,违规计数器不计数,只是交换的端口不转发违规计算机发送的数据。

18.4.3 粘滞端口安全的配置

本小节网络拓扑如图 18-1 所示,为了防止不合法的用户访问网络,要求配置粘滞端口安全,使交换机 S1 的 f0/3 端口只允许一台计算机接入,一旦端口有非法计算机接入,端口就进入限制安全违规模式。

步骤 1:在交换机 S1 的全局配置模式下启用端口安全,输入以下代码。

```
S1(config)#interface f0/3
S1(config-if)#switchport mode access
S1(config-if)#switchport port-security
```

步骤 2:配置粘滞端口安全,输入以下代码。

```
S1(config-if)#switchport port-security mac-address sticky
```

步骤 3:配置端口的安全违规模式,输入以下代码。

```
S1(config-if)#switchport port-security violation restrict
```

步骤 4:在交换机 S1 的特权执行模式下输入 show port-security address,查看交换机 S1 的安全 MAC 地址,如图 18-28 所示。

```
S1#show port-security address
        Secure Mac Address Table
-------------------------------------------------------------
Vlan    Mac Address       Type              Ports   Remaining Age
                                                      (mins)
----    -----------       ----              -----   -------------
 1      54be.f74e.2502    SecureSticky      Fa0/3      -
-------------------------------------------------------------
Total Addresses in System (excluding one mac per port)    : 0
Max Addresses limit in System (excluding one mac per port): 8192
```

图 18-28 交换机 S1 的安全 MAC 地址

步骤 5:在交换机 S1 的特权执行模式下输入 show port-security interface f0/3,查看交换机 S1 的 f0/3 端口的安全设置,如图 18-29 所示。

```
S1#show port-security interface f0/3
Port Security              : Enabled
Port Status                : Secure-up
Violation Mode             : Restrict
Aging Time                 : 0 mins
Aging Type                 : Absolute
SecureStatic Address Aging : Disabled
Maximum MAC Addresses      : 1
Total MAC Addresses        : 1
Configured MAC Addresses   : 0
Sticky MAC Addresses       : 1
Last Source Address:Vlan   : 54be.f74e.2502:1
Security Violation Count   : 0
```

图 18-29 交换机 S1 的 f0/3 端口的安全设置

步骤 6：在交换机 S1 的特权执行模式下输入 show run interface f0/3，查看交换机 S1 运行配置文件中 f0/3 端口的配置，如图 18-30 所示。

```
S1#show run interface f0/3
Building configuration...

Current configuration : 233 bytes
!
interface FastEthernet0/3
 switchport mode access
 switchport port-security
 switchport port-security violation restrict
 switchport port-security mac-address sticky
 switchport port-security mac-address sticky 54be.f74e.2502
end
```

图 18-30 交换机 S1 运行配置文件中 f0/3 端口的配置

步骤 7：在计算机 PC2 的命令行界面输入 ping 192.168.1.254，检验联通性，如图 18-31 所示。

```
C:\>ping 192.168.1.254

正在 Ping 192.168.1.254 具有 32 字节的数据:
来自 192.168.1.254 的回复: 字节=32 时间=2ms TTL=255
来自 192.168.1.254 的回复: 字节=32 时间<1ms TTL=255
来自 192.168.1.254 的回复: 字节=32 时间<1ms TTL=255
来自 192.168.1.254 的回复: 字节=32 时间<1ms TTL=255

192.168.1.254 的 Ping 统计信息:
    数据包: 已发送 = 4，已接收 = 4，丢失 = 0 (0% 丢失)，
往返行程的估计时间(以毫秒为单位):
    最短 = 0ms，最长 = 2ms，平均 = 0ms
```

图 18-31 从计算机 PC2 ping 路由器 R1 的 f0/0 端口

步骤 8：把一台集线器 HUB1 接入交换机 S1 的 f0/3 端口，在集线器 HUB1 上连接计算机 PC2 和 PC3，网络拓扑如图 18-32 所示。

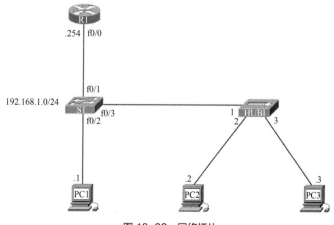

图 18-32 网络拓扑

当用集线器 HUB1 把计算机 PC3 接入交换机 S1 的 f0/3 端口时，显示图 18-33 所示的提示消息。

```
*Mar  1 00:12:23.398: %PORT_SECURITY-2-PSECURE_VIOLATION: Security violation occurred, caused by MAC address c85b.7684.c8
5d on port FastEthernet0/3.
*Mar  1 00:12:29.035: %PORT_SECURITY-2-PSECURE_VIOLATION: Security violation occurred, caused by MAC address c85b.7684.c8
5d on port FastEthernet0/3.
*Mar  1 00:12:35.377: %PORT_SECURITY-2-PSECURE_VIOLATION: Security violation occurred, caused by MAC address c85b.7684.c8
5d on port FastEthernet0/3.
```

图 18-33　提示消息

步骤 9：在交换机 S1 的特权执行模式下查看交换机 S1 的 f0/3 端口的状态，如图 18-34 所示。

```
S1#show interface f0/3 status

Port       Name               Status       Vlan       Duplex  Speed Type
Fa0/3                         connected    1          a-full  a-100 10/100BaseTX
```

图 18-34　交换机 S1 的 f0/3 端口的状态

步骤 10：在交换机 S1 的特权执行模式下输入 show port-security address，查看安全 MAC 地址，如图 18-35 所示。

```
S1#show port-security address
           Secure Mac Address Table
-----------------------------------------------------------------
Vlan    Mac Address       Type                     Ports   Remaining Age
                                                           (mins)
----    -----------       ----                     -----   -------------
1       54be.f74e.2502    SecureSticky             Fa0/3      -
-----------------------------------------------------------------
Total Addresses in System (excluding one mac per port)     : 0
Max  Addresses limit in System (excluding one mac per port) : 8192
```

图 18-35　交换机 S1 的安全 MAC 地址

步骤 11：在交换机 S1 的特权执行模式下输入 show port-security interface f0/3，查看 f0/3 端口的安全设置，如图 18-36 所示。

```
S1#show port-security interface f0/3
Port Security                : Enabled
Port Status                  : Secure-up
Violation Mode               : Restrict
Aging Time                   : 0 mins
Aging Type                   : Absolute
SecureStatic Address Aging   : Disabled
Maximum MAC Addresses        : 1
Total MAC Addresses          : 1
Configured MAC Addresses     : 0
Sticky MAC Addresses         : 1
Last Source Address:Vlan     : c85b.7684.c85d:1
Security Violation Count     : 31
```

图 18-36　交换机 S1 的 f0/3 端口的安全设置

步骤 12：在计算机 PC3 的命令行界面输入 ping 192.168.1.2，检验联通性，如图 18-37 所示。

```
C:\>ping 192.168.1.2

正在 Ping 192.168.1.2 具有 32 字节的数据：
来自 192.168.1.2 的回复: 字节=32 时间=1ms TTL=128
来自 192.168.1.2 的回复: 字节=32 时间<1ms TTL=128
来自 192.168.1.2 的回复: 字节=32 时间<1ms TTL=128
来自 192.168.1.2 的回复: 字节=32 时间<1ms TTL=128

192.168.1.2 的 Ping 统计信息：
    数据包: 已发送 = 4，已接收 = 4，丢失 = 0 (0% 丢失)，
往返行程的估计时间(以毫秒为单位)：
    最短 = 0ms，最长 = 1ms，平均 = 0ms
```

图 18-37　从计算机 PC3 ping 计算机 PC2

步骤 13：在计算机 PC3 的命令行界面输入 ping 192.168.1.254，检验联通性，如图 18-38 所示。

```
C:\>ping 192.168.1.254
正在 Ping 192.168.1.254 具有 32 字节的数据：
来自 192.168.1.3 的回复：无法访问目标主机。
来自 192.168.1.3 的回复：无法访问目标主机。
来自 192.168.1.3 的回复：无法访问目标主机。
来自 192.168.1.3 的回复：无法访问目标主机。

192.168.1.254 的 Ping 统计信息：
    数据包：已发送 = 4，已接收 = 4，丢失 = 0 (0% 丢失)，
```

图 18-38　从计算机 PC3 ping 路由器 R1 的 f0/0 接口

步骤 14：在计算机 PC2 的命令行界面输入 ping 192.168.1.254，检验联通性，如图 18-39 所示。

```
C:\>ping 192.168.1.254
正在 Ping 192.168.1.254 具有 32 字节的数据：
来自 192.168.1.254 的回复：字节=32 时间=1ms TTL=255
来自 192.168.1.254 的回复：字节=32 时间<1ms TTL=255
来自 192.168.1.254 的回复：字节=32 时间<1ms TTL=255
来自 192.168.1.254 的回复：字节=32 时间<1ms TTL=255

192.168.1.254 的 Ping 统计信息：
    数据包：已发送 = 4，已接收 = 4，丢失 = 0 (0% 丢失)，
往返行程的估计时间(以毫秒为单位)：
    最短 = 0ms，最长 = 1ms，平均 = 0ms
```

图 18-39　从计算机 PC2 ping 路由器 R1 的 f0/0 口

通过查看交换机 S1 运行配置文件中 f0/3 端口的配置可以发现，粘滞安全 MAC 地址会被写入交换机的运行配置文件。通过查看交换机 S1 的 f0/3 端口的状态和安全设置可以发现，当交换机 S1 的 f0/3 端口接入的计算机数量超过一台时，会出现端口的安全违规，端口进入限制安全违规模式，会显示提示信息，违规计数器会计数。通过查看从计算机 PC3 ping 计算机 PC2 返回的结果可以发现，计算机 PC2 和 PC3 已经实现了互通，说明计算机 PC3 已经正确接入网络。通过查看从计算机 PC2 和计算机 PC3 ping 路由器 R1 的 f0/0 接口的结果可以发现，计算机 PC2 能够通过交换机 S1 的 f0/3 端口访问路由器 R1 的 f0/0 接口，而计算机 PC3 无法通过交换机 S1 的 f0/3 端口访问路由器 R1 的 f0/0 接口。限制安全违规模式下会发送安全违规的通知，违规计数器会计数，交换机的端口不转发违规计算机发送的数据。

18.5　项目小结

本项目完成了端口安全的配置。当网络中的交换机启用了端口安全时，可以阻止接入端口的不合法用户访问网络。端口安全分为静态端口安全、动态端口安全和粘滞端口安全 3 种。静态端口安全使用管理员手动配置的静态安全 MAC 地址，动态端口安全使用交换机自己学习的动态安全 MAC 地址，粘滞端口安全使用交换机自己学习的粘滞安全 MAC 地址，粘滞安全 MAC 地址会被写入运行配置文件，如果保存运行配置文件，那么粘滞安全 MAC 地址会一直保存在配置文件中。端口的安全违规模式有保护安全违规模式、限制安全违规模式和关闭安全违规模式 3 种。保护安全违规模式下，端口不转发违规计算机发送的数据，不发送安全违规的通知，违规计数器也不计数；限制安全违规模式下，端口不转发违规计算机发送的数据，会发送安全违规的通知，违规计数器要计数；关闭安全违规模式下，端口处于 error-disabled 状态，会发送安全违规的通知，违规计数器也要计数，关闭安全违规模式是交换机端口默认的安全违规模式。

18.6 拓展训练

本项目的拓展训练网络拓扑如图 18-40 所示，要求完成如下配置：

（1）完成静态端口安全的配置，使交换机 S1 的 f0/1 端口仅允许计算机 PC1 接入。如果有其他计算机接入，那么端口将进入保护安全违规模式；

（2）完成动态端口安全的配置，使交换机 S1 的 f0/2 端口仅允许两台计算机接入。如果有两台以上的计算机接入，那么端口将进入关闭安全违规模式。

（3）完成粘滞端口安全的配置，使交换机 S1 的 f0/3 端口仅允许 3 台计算机接入。如果有 3 台以上的计算机接入，那么端口将进入限制安全违规模式。

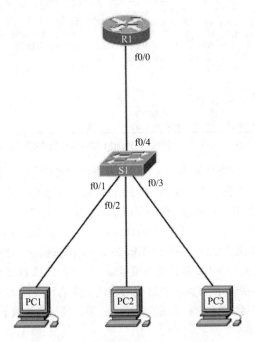

图 18-40　项目 18 拓展训练网络拓扑

模块六
路由器的管理

在运行过程中,网络设备的正常运行是保证网络正常运行的基础。网络设备的管理非常重要,熟练运用网络设备管理常用的技术也是网络工程技术人员需要掌握的基本技能。本模块以路由器的管理为例,主要介绍路由器的密码恢复、路由器 IOS 的备份与恢复等。

项目 19

路由器的密码恢复

19.1 用户需求

某学校网络中心由于原来的网络管理员离职，一台路由器的 Console 口登录密码和远程登录密码丢失。由于近期正在对网络进行升级和改造，需要完善这台设备的配置，但是由于密码丢失，无法登录设备，怎样解决这个问题？

19.2 知识梳理

19.2.1 配置寄存器

路由器的配置寄存器以一个十六进制数表示，配置寄存器就像计算机中用于控制启动过程的基本输入/输出系统（Basic Input/Output System，BIOS），BIOS 可以设置计算机将哪个存储器作为引导设备启动。

路由器密码的恢复

19.2.2 路由器密码的恢复

路由器的配置寄存器的默认值是 0x2102，表示启动时加载 NVRAM 的配置文件。路由器的密码丢失后，在启动时按组合键"Ctrl+Break"进入 ROMmon 模式，然后将配置寄存器的值修改为 0x2142，使路由器在启动时不加载 NVRAM 配置文件，这样路由器在启动时就可以绕过登录密码。当启动路由器后，可以手动将启动配置文件加载到运行配置文件，然后修改密码；Console 口登录密码和远程登录密码都修改完成后，把配置寄存器的值改回默认值。

19.3 方案设计

本项目中，路由器的 Console 口登录密码和远程登录密码都丢失了，这种情况下可以恢复密码。可以通过恢复密码给路由器设置新的 Console 口登录密码。如果密码是明文格式的，则可以通过查看配置文件读出密码，不需要修改；如果密码是密文格式的，则查看配置文件无法读出密码，这种情况需要修改密码。

19.4 项目实施

本项目中，路由器的 Console 口登录密码和远程登录密码都丢失了，要求完成路由器密码的恢复，使路由器能通过 Console 口或远程登录正常登录路由器。

步骤 1：路由器的 Console 口登录。

通过 Console 电缆把路由器的 Console 口和计算机的 COM 口连接起来，并在计算机上运行终端软件，进入路由器的用户执行模式。

步骤 2：重启路由器。

关闭路由器的电源开关，然后重新打开电源开关。

步骤 3：进入 ROMmon 模式。

在路由器启动过程的 60s 内按终端键盘上的"Ctrl+Break"组合键，使路由器进入 ROMmon 模式。当出现"rommon 1>"提示符时，说明已进入 ROMmon 模式。

> **注意** 在重启路由器时，如果按"Ctrl+Break"组合键路由器没有反应，那么可以把终端软件的 baud rate 参数的值改为 1200，这时在屏幕上不会输出任何信息，然后将路由器电源开关关闭，再重新打开电源开关，使路由器重新启动，在路由器启动过程的 60s 内按住空格键 10~15s，这个操作与按"Ctrl+Break"组合键的结果相似，然后把 baud rate 参数的值改为 9600，这时就可以进入 ROMmon 模式。

步骤 4：输入 confreg 0x2142 和 reset。

```
rommon 1 >confreg 0x2142
rommon 1 >reset
```

修改配置寄存器的值为 0x2142，使路由器在重新启动时不加载配置文件，从而绕过配置的密码。

步骤 5：当路由器进入初始配置模式时输入 no，或者按"Ctrl+C"组合键跳过初始配置过程。

步骤 6：在特权执行模式下输入 copy startup-config running-config。

```
Router#copy startup-config running-config
```

将 NVRAM 中保存的启动配置文件复制到内存中。注意，输入这条命令时，一定不能把 startup-config 和 running-config 的位置互换，一旦这两个参数的位置互换了，路由器的原有配置就会被删除。

步骤 7：在特权执行模式下输入 show running-config。

```
Router#show running-config
```

通过查看配置文件可以看到密文格式或明文格式的密码，明文格式的密码可以被直接读出来并继续使用，密文格式的密码就必须修改。

步骤 8：在全局配置模式下修改密码。

对密文格式的密码进行更改。

步骤 9：修改配置寄存器的值为默认值，在路由器的特权执行模式下输入以下代码。

```
Router(config)#config-register 0x2102
```

修改配置寄存器的值为 0x2102。

步骤 10：按"Ctrl+Z"组合键或输入 end 退出全局配置模式，进入特权执行模式。

步骤 11：保存配置。

19.5　项目小结

本项目完成了路由器的密码恢复。要实现路由器的密码恢复，需要先重启路由器，重启过程中，按"Ctrl+Break"组合键，使路由器进入 ROMmon 模式，在这个模式下修改配置寄存器的值，使路由器启动过程中不加载启动配置文件，从而绕过密码，直接进入用户执行模式，然后把配置文件手动加载到路由器的内存。这样，明文格式的密码通过查看配置文件就可以被直接读出来并继续使用，密文格式的密码需要修改，修改完成后，需要把配置寄存器的值再修改成默认值并保存配置。

19.6 拓展训练

某学校网络中，一台核心交换机的 Console 口登录和远程登录密码丢失，要求完成交换机的密码恢复，使交换机能通过 Console 口或远程登录正常登录。

项目 20
路由器IOS的备份与恢复

20.1 用户需求

某学校校园网中有一台路由器,为防止其 IOS 丢失,需要对 IOS 文件进行备份,以便在 IOS 不小心丢失时对 IOS 进行恢复,那么怎样对路由器的 IOS 进行备份与恢复呢?

20.2 知识梳理

路由器 IOS 的备份与恢复

20.2.1 IOS

IOS 是在思科网络设备上使用的操作系统,是一个将路由、交换、安全和其他网络互联技术集成到单个多任务操作系统中的软件包。大多数思科设备,无论其种类和大小如何,都离不开 IOS。类似于计算机的桌面操作系统,交换机或路由器的 IOS 为网络技术人员提供一个界面,技术人员可以输入命令配置设备或为设备编程,以实现各种网络功能。不同网络设备的 IOS 不尽相同,具体取决于设备的用途和支持的功能。使用何种 IOS 版本取决于使用的设备类型和所需的功能。当所有设备都有默认的 IOS 和功能集时,可以升级 IOS 版本或功能集以获得更多的功能。

20.2.2 IOS 的存放位置

IOS 文件的大小为几兆字节,它被存储在闪存中。闪存可提供非易失性存储,设备断电时,闪存中的内容不会丢失。必要时可以更改或覆盖闪存内容,这样可直接将 IOS 升级为新版本或添加新功能,而无须更换硬件。此外,闪存可以同时存储多个 IOS 版本。

许多思科设备启动时,IOS 从闪存被复制到 RAM。RAM 具有许多功能,包括存储设备用于支持网络运营的数据。设备工作时,IOS 在 RAM 中运行。由于重新通电时 RAM 中的数据会丢失,因此 RAM 又被称为易失性存储器。

特定版本的 IOS 所需的闪存和 RAM 容量差别很大。为了便于维护和规划网络,确定每个设备运行的 IOS 版本需要的闪存和 RAM 非常重要。某些 IOS 版本对闪存和 RAM 的要求可能超过某些设备上可以安装的闪存和 RAM,在设备升级 IOS 前,应检查需要升级 IOS 设备的闪存和 RAM 容量是否满足新版本 IOS 需要的容量。

20.2.3 TFTP

简单文件传送协议（Trivial File Transfer Protocol，TFTP）是 TCP/IP 协议族中的一个协议，用来在客户端与服务器之间进行简单文件传送，提供不复杂、开销不大的文件传输服务。为了防止路由器等设备的系统映像及配置文件损坏或被意外删除，网络 TFTP 服务器保留 IOS 映像或者配置文件的备份。网络 TFTP 服务器可以是一台路由器、工作站，也可以是主机系统。IOS 映像和配置文件可通过网络上传和下载。

20.2.4 Xmodem 协议

Xmodem 协议是一种在串口通信中被广泛应用的异步文件传输协议，分为标准 Xmodem 和 1K-Xmodem 两种，前者以 128 字节块的形式传输数据，后者的字节块为 1024 字节，并且对每个块都使用一个校验过程来进行错误检测。在校验过程中，如果接收方一个块的校验和与发送方的校验和相同，那么接收方就向发送方发送一个确认字节（ACK）。由于 Xmodem 协议需要对每个块都进行认可，会导致性能有所下降，特别是延时比较长的场合，这种协议显得效率较低。Ymodem 协议是 Xmodem 协议的改进版，具有传输快速、稳定的优点。Ymodem 协议支持一次传输 1024 字节，还支持同时传输多个文件。

20.2.5 配置命令

1. 备份 IOS 文件并上传到 TFTP 服务器

```
Router#copy flash: tftp:
```

按"Enter"键后，设备会提示输入源文件名、远程主机（TFTP 服务器）的地址或者名称和目标文件名。

2. 从 TFTP 服务器恢复设备的 IOS

```
Router#copy tftp: flash:
```

按"Enter"键后，设备会提示输入远程主机（TFTP 服务器）的地址或者名称、源文件名和目标文件名。

3. 在 ROMmon 模式下使用 Xmodem 协议恢复 IOS

```
rommon 1> xmodem [-cyr] [filename]
```

-cyr：此选项根据配置参数的不同有不同的含义，-c 表示 CRC-16，y 表示 Ymodem 协议，r 表示将映像复制到 RAM 中。

filename：需要传输的文件的名称。

4. 指定 Flash 作为思科 IOS 映像的源设备

```
Router(config)#boot system flash0://filename
```

5. 指定从 ROM 引导

```
Router(config)#boot system rom
```

指定 TFTP 服务器作为思科 IOS 映像的一个来源，如果找不到正确的 IOS，则可以指定从 ROM 引导。

20.3 方案设计

本项目中，某学校网络有一台路由器的 IOS 丢失了，需要管理员对路由器的 IOS 进行恢复。如果之前做过路由器 IOS 文件的备份，则可以用备份的 IOS 文件完成路由器 IOS 的恢复。在网络管理过程中，为了便于网络设备的管理及后期的维护，一般都会对设备的 IOS 文件进行备份。备份 IOS 文件可以借助 TFTP 服务器，通过 copy 命令把 IOS 文件上传到 TFTP 服务器。如果路由器的 IOS 丢失后，路由器没有重启仍然正常运行，那么这时可以借助 TFTP 服务器把 IOS 文件下载到路由器完成恢复；如果设备的 IOS 丢失后，设备已经重启了，那么需要通过 ROMmon 模式，用 TFTP 服务器恢复或者使用 Xmodem 协议恢复。

20.4 项目实施

20.4.1 备份 IOS 到 TFTP 服务器

某学校刚刚完成了网络的升级和改造，为了便于网络设备的管理及后期的维护，要求管理员对网络设备的 IOS 进行备份。

步骤 1：在管理员的计算机上运行 TFTP 服务器软件（TFTP 服务器软件有很多，本项目选择的是 Cisco TFTP Server），运行界面如图 20-1 所示。

图 20-1 Cisco TFTP Server 运行界面

步骤 2：单击 按钮，弹出"选项"对话框，如图 20-2 所示。

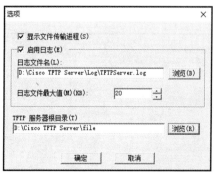

图 20-2 "选项"对话框

步骤 3：在"选项"对话框中，单击"日志文件名"文本框后面的"浏览"按钮，设置日志文件的存放位置；单击"TFTP 服务器根目录"文本框后面的"浏览"按钮，设置 TFTP 服务器接收的文件的存放位置。

步骤 4：按照图 20-3 所示的网络拓扑把计算机和路由器连接起来。

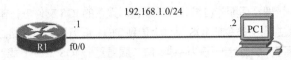

图 20-3 网络拓扑

步骤 5：配置路由器的 f0/0 接口和计算机的 IP 地址。

步骤 6：通过 Console 口登录路由器并进入特权执行模式，输入 dir，查看路由器闪存中的文件，如图 20-4 所示。

```
R1#dir
Directory of flash:/

    1  -rw-    23174468  Jun 2 2010 06:42:42 +00:00  c2801-ipbase-mz.124-15.T13.bin
    2  -rw-        2898  Jun 2 2010 06:44:10 +00:00  cpconfig-2801.cfg
    3  -rw-     2938880  Jun 2 2010 06:44:36 +00:00  cpexpress.tar
    4  -rw-        1038  Jun 2 2010 06:44:50 +00:00  home.shtml
    5  -rw-      122880  Jun 2 2010 06:45:04 +00:00  home.tar
    6  -rw-      527849  Jun 2 2010 06:45:20 +00:00  128MB.sdf
    7  -rw-     1697952  Jun 2 2010 06:45:48 +00:00  securedesktop-ios-3.1.1.45-k9.pkg
    8  -rw-      415956  Jun 2 2010 06:46:06 +00:00  sslclient-win-1.1.4.176.pkg

128704512 bytes total (99811328 bytes free)
```

图 20-4 查看路由器闪存中的文件

通过查看路由器闪存中的文件，找到路由器的 IOS 文件（即 c2801-ipbase-mz.124-15.T13.bin）。

步骤 7：备份 IOS 文件，在路由器 R1 的特权执行模式下，用复制命令把 IOS 文件上传到 TFTP 服务器，输入以下代码。

```
R1#copy flash:c2801-ipbase-mz.124-15.T13.bin tftp:
```

按"Enter"键后，会提示输入远程主机的地址或者主机名，输入 TFTP 服务器的地址 192.168.1.1 后，提示确认目标文件名（TFTP 服务器上的 IOS 镜像文件的名称），可以直接按"Enter"键，确认目标文件名并开始上传，当出现图 20-5 所示的内容时说明备份成功。

```
R1#copy flash:c2801-ipbase-mz.124-15.T13.bin tftp:
Address or name of remote host []? 192.168.1.1
Destination filename [c2801-ipbase-mz.124-15.T13.bin]?
.!!!!!!!!!!!!!!!!!!!!!!!!!!!!!!!!!!!!!!!!!!!!!!!!!!!!!!!!!!!!!!!!!!!!!!!!!!!!!!
23174468 bytes copied in 126.188 secs (183650 bytes/sec)
```

图 20-5 把路由器的 IOS 文件成功备份到 TFTP 服务器

步骤 8：进入 TFTP 服务器的文件存放位置 D:\Cisco TFTP Server\file，查看 IOS 文件，如图 20-6 所示。

图 20-6 查看 TFTP 服务器上的 IOS 文件

注意：在 IOS 文件备份过程中，一定要保证 TFTP 服务器软件是开启的。

20.4.2 用 TFTP 服务器恢复路由器的 IOS

某学校校园网中有一台正在运行的路由器，管理员发现该路由器的 IOS 丢失了，但设备没有重启，要求通过备份 TFTP 服务器上的 IOS 文件对路由器的 IOS 进行恢复。

如果路由器的 IOS 映像文件被删除或损坏，而设备没有重新启动，这是因为路由器的 IOS 正在路由器的 RAM 中运行，所以路由器仍然可以正常运行，此时不能重新启动路由器。如果重启路由器，那么将无法在闪存中找到有效的 IOS，路由器无法完成正常启动。管理员可以将备份在 TFTP 服务器中的 IOS 映像文件复制到路由器，完成路由器 IOS 的恢复。

步骤 1：按照图 20-3 所示的网络拓扑连接路由器和 TFTP 服务器，TFTP 服务器的配置步骤与 20.4.3 小节中的相同。

步骤 2：当 TFTP 服务器配置完成并与路由器建立连接后，在路由器的特权执行模式下，从 TFTP 服务器复制 IOS 文件，输入以下代码。

```
R1#copy tftp: flash:
```

按"Enter"键后，会提示输入远程主机的地址或者主机名，输入 TFTP 服务器的地址 192.168.1.1 后，提示确认源文件名，输入保存在 TFTP 服务器上的 IOS 镜像文件的名称后，提示确认目标文件名，可以直接按"Enter"键，确认目标文件名并开始从 TFTP 服务器复制，当出现如图 20-7 所示的内容时说明恢复成功。

```
R1#copy tftp: flash:
Address or name of remote host []? 192.168.1.1
Source filename []? c2801-ipbase-mz.124-15.T13.bin
Destination filename [c2801-ipbase-mz.124-15.T13.bin]?
Accessing tftp://192.168.1.1/c2801-ipbase-mz.124-15.T13.bin...
Loading c2801-ipbase-mz.124-15.T13.bin from 192.168.1.1 (via FastEthernet0/0): !!!!!!!!!!!!!!!!!!!!!!!!!!!!!!!!!!!!!!!!!!
!!!!!!!!!!!!!!!!!!!!!!!!!!!!!!!!!!!!!!!!!!
[OK - 23174468 bytes]

23174468 bytes copied in 94.184 secs (246055 bytes/sec)
```

图 20-7 成功从 TFTP 服务器恢复 IOS 到路由器

20.4.3 使用 ROMmon 模式恢复 IOS

某学校校园网中有一台路由器，管理员发现该路由器的 IOS 丢失了，设备已经重启但无法正常启动，要求通过备份 TFTP 服务器上的 IOS 映像文件恢复设备的 IOS。

如果路由器上的 IOS 被意外地从闪存中删除，并且路由器已经重新启动，那么路由器将无法加载 IOS，它会根据默认设置加载 ROMmon 提示符，进入 ROMmon 模式，如图 20-8 所示。

```
System Bootstrap, Version 12.4(13r)T, RELEASE SOFTWARE (fc1)
Technical Support: http://www.cisco.com/techsupport
Copyright (c) 2006 by cisco Systems, Inc.
PLD version 0x10
GIO ASIC version 0x127
c2801 platform with 393216 Kbytes of main memory
Main memory is configured to 64 bit mode with parity disabled

Readonly ROMMON initialized
program load complete, entry point: 0x8000f000, size: 0xcb80
program load complete, entry point: 0x8000f000, size: 0xcb80
loadprog: bad file magic number:     0x0
boot: cannot load "flash:"

System Bootstrap, Version 12.4(13r)T, RELEASE SOFTWARE (fc1)
Technical Support: http://www.cisco.com/techsupport
Copyright (c) 2006 by cisco Systems, Inc.
PLD version 0x10
GIO ASIC version 0x127
c2801 platform with 393216 Kbytes of main memory
Main memory is configured to 64 bit mode with parity disabled

Readonly ROMMON initialized
rommon 1 >
```

图 20-8 路由器进入 ROMmon 模式

要恢复设备的 IOS，可以使用 TFTP 服务器或者使用 Xmodem 协议两种方式。

1. 使用 TFTP 服务器恢复 IOS 映像

步骤 1：将计算机连接到丢失 IOS 的路由器的 Console 口。将 TFTP 服务器连接到该路由器的第一个以太网接口，在计算机上运行终端软件，实现计算机对路由器的 Console 口登录，并完成计算机 IP 地址的配置。

步骤 2：设置路由器 R1 的 ROMmon 变量，使路由器能够连接到 TFTP 服务器，输入以下代码。

```
rommon 1> IP_ADDRESS=192.168.1.2
rommon 2> IP_SUBNET_MASK=255.255.255.0
rommon 3> DEFAULT_GATEWAY=192.168.1.1
rommon 4> TFTP_SERVER=192.168.1.1
rommon 5> TFTP_FILE=c2801-ipbase-mz.124-15.T13.bin
```

> **注意** 输入 ROMmon 变量时，变量名要区分大小写；在"="的前后不能加入任何空格；为了正确输入，可以使用文本编辑器编辑变量，然后将完成编辑的变量复制并粘贴至终端窗口中。

步骤 3：在 ROMmon 提示符后输入 tftpdnld 命令。

```
rommon 6 > tftpdnld
```

步骤 4：按"Enter"键后，将显示所需的环境变量，并警告闪存中的所有现有数据都将被删除，如图 20-9 所示。

```
rommon 6 > tftpdnld
            IP_ADDRESS: 192.168.1.2
        IP_SUBNET_MASK: 255.255.255.0
       DEFAULT_GATEWAY: 192.168.1.1
           TFTP_SERVER: 192.168.1.1
             TFTP_FILE: c2801-ipbase-mz.124-15.T13.bin
          TFTP_MACADDR: 28:93:fe:5b:b0:4e
          TFTP_VERBOSE: Progress
      TFTP_RETRY_COUNT: 18
          TFTP_TIMEOUT: 7200
         TFTP_CHECKSUM: Yes
               FE_PORT: 0
         FE_SPEED_MODE: Auto Detect

Invoke this command for disaster recovery only.
WARNING: all existing data in all partitions on flash: will be lost!
Do you wish to continue? y/n:  [n]:
```

图 20-9 显示环境变量并警告闪存中的所有现有数据都将被删除

步骤 5：输入 y，然后按"Enter"键，路由器将尝试连接 TFTP 服务器，并开始下载。连接成功后，下载将开始，"!"会指示这一过程，每个"!"都表明路由器收到一个 UDP 数据包，如图 20-10 所示。

```
!!!!!!!!!!!!!!!!!!!!!!!!!!!!!!!!!!!!!!!!!!!!!!!!!!!!!!!!!!!!!!!!!!!!!!!!!!!!!!!!
File reception completed.
Validating checksum.
Copying file c2801-ipbase-mz.124-15.T13.bin to flash:.
program load complete, entry point: 0x8000f000, size: 0xcb80

Format: Drive communication & 1st Sector Write OK...
Writing Monlib sectors.
..............................................................................
Monlib write complete

Format: All system sectors written. OK...
Format: Operation completed successfully.

Format of flash: complete
program load complete, entry point: 0x8000f000, size: 0xcb80

rommon 7 >
```

图 20-10 使用 TFTP 服务器恢复 IOS 映像

步骤 6：以新的 IOS 映像重新加载路由器，输入以下代码。

```
rommon 7 > reset
```

按"Enter"键后，路由器将重启并加载新的 IOS 映像。

2. 使用 Xmodem 协议恢复 IOS 映像

路由器的 ROMmon 模式支持 Xmodem 协议，路由器能与计算机上的终端软件（如 SecureCRT）通信。

步骤 1：在路由器的 ROMmon 模式下修改波特率，输入以下代码。

```
rommon 1 > confreg
```

提示是否更改配置，如图 20-11 所示。

```
rommon 1 > confreg

            Configuration Summary
    (Virtual Configuration Register: 0xa102)
enabled are:
diagnostic mode
load rom after netboot fails
console baud: 9600
boot: image specified by the boot system commands
      or default to: cisco2-c2801

do you wish to change the configuration? y/n  [n]:
```

图 20-11　提示是否更改配置

步骤 2：输入 y，进入具体修改选项，一直输入 n，直到出现"change console baud rate? y/n"，输入 y，再输入 7，选择波特率为 115200，如图 20-12 所示。

```
do you wish to change the configuration? y/n  [n]: y
disable "diagnostic mode"? y/n  [n]: n
enable  "use net in IP bcast address"? y/n  [n]: n
disable "load rom after netboot fails"? y/n  [n]: n
enable  "use all zero broadcast"? y/n  [n]: n
enable  "break/abort has effect"? y/n  [n]: n
enable  "ignore system config info"? y/n  [n]: n
change console baud rate? y/n  [n]: y
0=9600, 1=4800, 2=1200, 3=2400, 4=19200, 5=38400, 6=57600, 7=115200
enter rate  [0]:
```

图 20-12　修改选项（1）

步骤 3：在出现的"change the boot characteristics? y/n"选项后，输入 n，在出现的"do you wish to change the configuration? y/n"选项后，输入 n，提示设备需要重启，如图 20-13 所示。

```
change console baud rate? y/n  [n]: y
0=9600, 1=4800, 2=1200, 3=2400, 4=19200, 5=38400, 6=57600, 7=115200
enter rate  [0]: 7
change the boot characteristics? y/n  [n]: n

            Configuration Summary
    (Virtual Configuration Register: 0xb922)
enabled are:
diagnostic mode
load rom after netboot fails
console baud: 115200
boot: image specified by the boot system commands
      or default to: cisco2-c2801

do you wish to change the configuration? y/n  [n]: n

You must reset or power cycle for new config to take effect
rommon 2 >
```

图 20-13　修改选项（2）

步骤 4：重新启动设备，输入以下代码。

```
rommon 2 > reset
```

步骤 5：将计算机通过 Console 电缆连接到路由器的 Console 口。管理员通过 Console 口登录路由器，登录时需将波特率设置为 115200。

步骤 6：在 ROMmon 提示符后输入 xmodem 命令。

```
rommon 3 > xmodem -c c2801-ipbase-mz.124-15.T13.bin
```

按"Enter"键后，会出现如图 20-14 所示的提示消息。

```
rommon 1 > xmodem -c c2801-ipbase-mz.124-15.T13.bin
Do not start the sending program yet...
program load complete, entry point: 0x8000f000, size: 0xcb80
        File size      Checksum     File name
WARNING: All existing data in flash will be lost!
Invoke this application only for disaster recovery.
Do you wish to continue? y/n  [n]:
```

图 20-14　提示消息

步骤 7：输入 y，接受出现的所有提示消息并继续执行，如图 20-15 所示。

```
rommon 1 > xmodem -c c2801-ipbase-mz.124-15.T13.bin
Do not start the sending program yet...
program load complete, entry point: 0x8000f000, size: 0xcb80
        File size      Checksum     File name
WARNING: All existing data in flash will be lost!
Invoke this application only for disaster recovery.
Do you wish to continue? y/n  [n]:  y
Ready to receive file b ...
```

图 20-15　接受提示消息并继续执行

步骤 8：使用 SecureCRT 发送文件，单击菜单栏中的"Transfer"菜单，在弹出的下拉菜单中选择"Send Xmodem"命令，如图 20-16 所示。

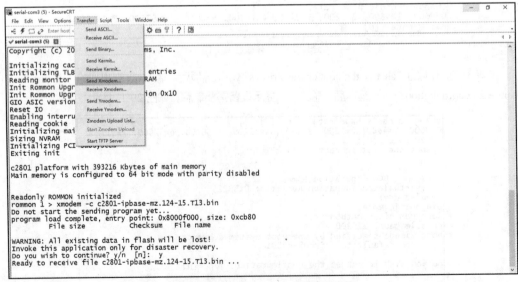

图 20-16　使用 Xmodem 协议发送文件

此时弹出"Select File to Send using Xmodem"对话框，如图 20-17 所示。

项目 20
路由器 IOS 的备份与恢复

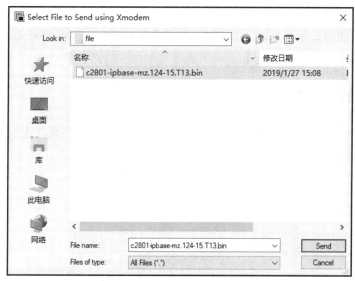

图 20-17 "Select File to Send using Xmodem" 对话框

步骤 9：在"Select File to Send using Xmodem"对话框的"Look in"下拉列表中选择要传输的 IOS 映像所在的位置，在"File name"下拉列表中选择要传输的 IOS 映像文件的名称，单击"Send"按钮，随后将出现一个显示下载状态的界面，如图 20-18 所示。主机和路由器需要经过几秒之后才会开始传输信息。

```
WARNING: All existing data in flash will be lost!
Invoke this application only for disaster recovery.
Do you wish to continue? y/n  [n]:  y
Ready to receive file c2801-ipbase-mz.124-15.T13.bin ...

Transferring c2801-ipbase-mz.124-15.T13.bin, errors 2...
   2%      565 KB        1 KB/sec      04:17:06 ETA    2 Errors
```

图 20-18 显示下载状态的界面

下载开始后，Packet（数据包）和 Elapsed（已用）字段的值将会增加，等待传输完成。

使用 Xmodem 协议恢复路由器的 IOS 的方式与使用 TFTP 服务器相比速度慢很多，但是这种方式使用 Console 电缆完成文件传输，不用搭建 TFTP 服务器。

20.5 项目小结

本项目完成了路由器 IOS 的备份与恢复，路由器的 IOS 是路由器的操作系统，如果路由器中的 IOS 被误删或者丢失，那么设备将无法正常运行。在实际的网络运维中，为了便于网络管理，需要备份设备的 IOS 和配置文件，可以把设备的 IOS 和配置文件通过 copy 命令备份到 TFTP 服务器。当路由器的 IOS 丢失后，如果设备没有重启，则可以直接从 TFTP 服务器下载 IOS 到路由器的闪存中，完成路由器 IOS 的恢复。如果路由器的 IOS 丢失了，并且路由器也重启了，但路由器无法正常启动，那么此时会进入 ROMmon 模式，可以使用 TFTP 服务器或者 Xmodem 协议两种方式恢复 IOS。使用 TFTP 服务器恢复 IOS 的速度相对比较快，使用 Xmodem 协议的速度会比较慢，把波特率修改为 115200，会提高传输速度。如果要升级设备的 IOS，那么只要把路由器的原有 IOS 从闪存中删掉，把新版本 IOS 复制到路由器的闪存中并重启设备即可。升级时需要核对设备的闪存和 RAM 容量能否满足新版本 IOS 的要求。如果设备的闪存容量可以存放多个 IOS，那么升级时也可以保留原有的 IOS 文件，直接把 IOS 的新版本复制到路由器，然后用 boot system 命令指定路由器用于引导设备的 IOS 文件。

20.6 拓展训练

某学校网络中，一台交换机的 IOS 丢失，并且设备已经重新启动，要求完成交换机的 IOS 的恢复与升级，并写出 IOS 的升级步骤。